Heritable
HUMAN GENOME
Editing

International Commission on the
Clinical Use of Human Germline Genome Editing

A Consensus Study Report of

NATIONAL ACADEMY OF MEDICINE
NATIONAL ACADEMY OF SCIENCES

and

THE
ROYAL
SOCIETY

THE NATIONAL ACADEMIES PRESS
Washington, DC
www.nap.edu

THE NATIONAL ACADEMIES PRESS 500 Fifth Street, NW Washington, DC 20001

This activity was supported by Contract No. HHSN263201800029I / Order No. 75N98019F00852 from the U.S. National Institutes of Health, Grant No. 2019 HTH 009 from the Rockefeller Foundation, and Grant 218375/Z/19/Z from the Wellcome Trust, with additional support from the Royal Society of the United Kingdom, the Cicerone Endowment Fund of the U.S. National Academy of Sciences, and the NAM Initiatives Fund of the U.S. National Academy of Medicine. Any opinions, findings, conclusions, or recommendations expressed in this publication do not necessarily reflect the views of any organization or agency that provided support for the project.

International Standard Book Number-13: 978-0-309-67113-2
International Standard Book Number-10: 0-309-67113-2
Digital Object Identifier: https://doi.org/10.17226/25665
Library of Congress Control Number: 2020949045

Additional copies of this publication are available from the National Academies Press, 500 Fifth Street, NW, Keck 360, Washington, DC 20001; (800) 624-6242 or (202) 334-3313; http://www.nap.edu.

Printed in the United States of America

Suggested citation: National Academy of Medicine, National Academy of Sciences, and the Royal Society. 2020. *Heritable Human Genome Editing*. Washington, DC: The National Academies Press. https://doi.org/10.17226/25665.

INTERNATIONAL COMMISSION ON THE CLINICAL USE OF HUMAN GERMLINE GENOME EDITING

HAOYI WANG, Ph.D., Professor, State Key Laboratory of Stem Cell and Reproductive Biology, Institute of Zoology, Institute for Stem Cell and Regeneration, Chinese Academy of Sciences, China

ANNA WEDELL, M.D., Ph.D., Professor and Head of Clinic, Centre for Inherited Metabolic Diseases, Karolinska Institute and Karolinska University Hospital, Sweden

Study Staff

KATHERINE W. BOWMAN, Senior Program Officer (Commission Co-Director), NAM/NAS

JONNY HAZELL, Senior Policy Adviser (Commission Co-Director), The Royal Society

MEGHAN ANGE-STARK, Associate Program Officer, NAM/NAS (from January 2020)

SARAH BEACHY, Senior Program Officer, NAM/NAS

CONNIE BURDGE, Policy Adviser, The Royal Society

DEBBIE HOWES, Personal Assistant and Events Coordinator, The Royal Society

STEVEN KENDALL, Program Officer, NAM/NAS

DAVID KUNTIN, Intern (April-July 2019), The Royal Society

DOMINIC LoBUGLIO, Senior Program Assistant, NAM/NAS (from November 2019)

ROB QUINLAN, Intern (July-September 2019), The Royal Society

FLANNERY WASSON, Senior Program Assistant, NAM/NAS (until November 2019)

ISABEL WILKINSON, Intern (September-December 2019), The Royal Society

EMMA WOODS, Head of Policy, The Royal Society

Consultant

STEVE OLSON, Writer

Special Acknowledgments

We are immensely grateful for the dedication and talent brought to bear on the work of this Commission by our fellow commissioners, the exceptional staff members of the NAS and the Royal Society, the external reviewers and monitors, and the many scientists, clinicians, and people with inherited disorders who have generously lent their time and insight to this project. All have far exceeded what we ever could have asked of them. Finally, we thank the members of the International Oversight Board for this study, who ensured that our report underwent a rigorous process of information gathering and external review prior to publication.

With gratitude and admiration,

Kay E. Davies, D.Phil. (Co-chair)
Richard P. Lifton, M.D., Ph.D. (Co-chair)

Foreword

The appointment of this Commission and the beginning of its deliberations occurred at a time when the reported birth of the "CRISPR babies" in China was fresh in many minds. This event made clear the absence of broad international consensus regarding both the societal acceptability of particular applications of heritable human genome editing (HHGE) and the scientific evidence that would be needed to demonstrate that HHGE could be done safely.

It was recognized that, without evidence of high efficiency and specificity to ensure that only the desired changes were introduced into the genome, there was continuing risk of ad hoc editing efforts that could cause significant harm to individuals. Moreover, given that heritable changes would be introduced that could be passed to subsequent generations, it was clear that careful consideration would need to be given to the specific applications of the editing technology.

During the preparation of this report, pressing issues have intervened. With the emergence of the SARS-CoV-2 coronavirus, the world's attention has been focused on the health, economic, and social consequences of the devastating COVID-19 pandemic, including the social inequalities of its impact in many countries. With intense protests that have taken place in many countries, the world's attention has also been focused on calls for changes to address racial injustice and inequities. These twin upheavals have underscored that we live in an interconnected world, where what happens in one country touches all countries, and that science occurs in a societal context. Although of a very different nature, the potential use of HHGE is an issue that transcends individual countries, deserves wide-ranging global discussions, and entails important issues of equity.

Genetic diseases can impose a major burden on families. For many prospective parents, viable options for having genetically-related, unaffected children are already available; but for others, due to genetics or reduced fertility, current alternatives may never be successful. HHGE might, in the future, provide a reproductive option for such couples.

At the same time, it is important to recognize that the idea of making intentional modifications to the human germline evokes to some the eugenics movements of the late 19th century and first half of the 20th century, which promoted now-discredited theories that led to the persecution of whole groups, based on race, religion, class, and ability. Should any nation decide to permit HHGE, it is vitally important that bias and discrimination be avoided. In addition, there must be constraints that prevent the use of HHGE for cases that are not medically justified interventions and not based on a rigorous understanding of genetics.

Great caution must also be taken in the development of genetic technologies like HHGE, fundamentally because of the personal and social contexts and broader societal and ethical issues that surround their application. Proposed uses of these technologies must reflect the conditions and needs of diverse human populations around the world. They should be deployed in ways that prevent harm and ensure equitable access to their benefits. The technologies themselves and the rigorous oversight structures established to regulate their use should be developed in ways that respect the human rights and inherent dignity of all persons.

The Commission is concerned that both the development and use of HHGE and allied assisted reproductive technologies (ARTs) must be properly regulated and overseen. In particular, it is important to avoid irresponsible practices in the use of HHGE. In making its recommendations, this Commission has taken into account the unfortunate fact that the practice of ART around the world too often lacks appropriate oversight.

Matters of equitable access are of course also raised by other ARTs and by health care in general, but these issues deserve note here. There is no doubt that the economic costs of developing and using the technology will be substantial. Moreover, since there are already viable alternatives for prospective parents to have genetically-related, unaffected offspring in the vast majority of cases, the benefits will accrue to very few prospective parents. Nonetheless, it is possible that HHGE might someday become sufficiently safe, robust, and efficient to be routinely applied in conjunction with ART to provide an improved option that would reduce the burden to women of repeated cycles of ovarian stimulation. Equitable access is the province of national jurisdictions, and the Commission recognizes the cost of development and the breadth of access to be issues that must be considered.

The Commission was specifically tasked with defining a responsible pathway for clinical use of HHGE, should a decision be made by any nation to permit its use. In fulfilling this assignment, we have considered current understanding in the areas of human genetics, genome editing, reproductive technologies, and associated social and ethical issues. This report is the product of our deliberations.

International Commission on the Clinical
Use of Human Germline Genome Editing

Contents

xiii

Boxes and Figures

BOXES

FIGURES

Summary

Rapidly advancing technical capabilities in genome editing, and the reported use of heritable human genome editing (HHGE) in 2018 leading to the birth of children whose DNA had been edited, led to renewed global calls for consideration of the scientific, societal, and governance issues associated with this technology. The possibility of heritable editing occurs when alterations to genomic DNA are made in gametes (eggs or sperm) or any cells that give rise to gametes, including the single cell zygote resulting from fertilization of an egg by a sperm cell, or cells of an early embryo. Changes made to the genetic material in such cells can be passed on to subsequent generations.

No country has yet decided that it would be appropriate to move forward clinically with HHGE, and clinical use of the technology is currently explicitly prohibited or not explicitly regulated in many countries. HHGE could represent an important option for prospective parents with a known risk of transmitting a genetic disease to have a genetically-related child without that disease and its associated morbidity and mortality. However, it will be essential to establish safe and effective methodologies that could form the necessary steps in a translational pathway for any clinical uses of HHGE. Assuming the existence of a safe and effective methodology, the decision to permit the clinical use of HHGE and, if so, for which specific applications, must ultimately rest with individual countries following informed societal debate of both ethical and scientific considerations.

This societal debate would include a range of issues and questions raised by HHGE, as well as how it might address important unmet needs within a country, informed by the views of patients and their families; ethical, moral, and religious views; potential long-term societal implications; and issues of cost and access. The societal considerations are the subject of ongoing national and international conversations, including current work by the WHOs Expert Advisory Committee on Developing Global Standards for Governance and Oversight of Human Genome Editing, which is deliberating on national and global governance.

The International Commission on the Clinical Use of Human Germline Genome Editing, which was convened by the U.S. National Academy of Medicine, the U.S. National Academy of Sciences, and the U.K.'s Royal Society and includes members from 10 countries, was tasked with addressing the scientific considerations that would be needed to inform broader societal decision making. This task involves considering technical, scientific, medical, and regulatory requirements, as well as those societal and ethical issues that are inextricably linked to these requirements, such as the significance of uncertainties related to outcomes, and potential benefits and harms to participants in clinical uses of HHGE.

This report does not make judgments about whether any clinical uses of a *safe and effective* HHGE methodology, if established by preclinical research, should at some point be permitted. The report instead seeks to determine whether the safety and efficacy of genome editing methodologies and associated assisted reproductive technologies (ARTs) are or could be sufficiently well developed to permit responsible clinical use of HHGE; identifies initial potential applications of HHGE for which a responsible clinical translational pathway can currently be defined; and delineates the necessary elements of such a translational pathway. It also elaborates national and international mechanisms necessary for appropriate *scientific* governance of HHGE, while recognizing that additional governance mechanisms may be needed to address societal considerations that lie beyond the Commission's charge. Box S-1 provides the full set of report recommendations; the Summary text provides the context for these.

CURRENT STATE OF SCIENTIFIC UNDERSTANDING

To assess what would be needed for a responsible translational pathway toward HHGE requires evaluating the state of scientific understanding of the effects of making genetic changes and of the procedures necessary to perform and to characterize the results of genome editing in human germline cells and embryos.

BOX S-1
Report Recommendations

Recommendation 1: No attempt to establish a pregnancy with a human embryo that has undergone genome editing should proceed unless and until it has been clearly established that it is possible to efficiently and reliably make precise genomic changes without undesired changes in human embryos. These criteria have not yet been met, and further research and review would be necessary to meet them.

Recommendation 2: Extensive societal dialogue should be undertaken before a country makes a decision on whether to permit clinical use of heritable human genome editing (HHGE). The clinical use of HHGE raises not only scientific and medical considerations but also societal and ethical issues that were beyond the Commission's charge.

Recommendation 3: It is not possible to define a responsible translational pathway applicable across all possible uses of heritable human genome editing (HHGE) because the uses, circumstances, and considerations differ widely, as do the advances in fundamental knowledge that would be needed before different types of uses could be considered feasible.

Clinical use of HHGE should proceed incrementally. At all times, there should be clear thresholds on permitted uses, based on whether a responsible translational pathway can be and has been clearly defined for evaluating the safety and efficacy of the use, and whether a country has decided to permit the use.

Recommendation 4: Initial uses of heritable human genome editing (HHGE), should a country decide to permit them, should be limited to circumstances that meet all of the following criteria:
1. the use of HHGE is limited to serious monogenic diseases; the Commission defines a serious monogenic disease as one that causes severe morbidity or premature death;
2. the use of HHGE is limited to changing a pathogenic genetic variant known to be responsible for the serious monogenic disease to a sequence that is common in the relevant population and that is known not to be disease-causing;
3. no embryos without the disease-causing genotype will be subjected to the process of genome editing and transfer, to ensure that no individuals resulting from edited embryos were exposed to risks of HHGE without any potential benefit; and
4. the use of HHGE is limited to situations in which prospective parents (i) have no option for having a genetically-related child that does not have the serious monogenic disease, because none of their embryos would be genetically unaffected in the absence of genome editing; or (ii) have extremely poor options, because the expected proportion of unaffected embryos would be unusually low, which the Commission defines as

continued

BOX S-1 Continued

25 percent or less, and have attempted at least one cycle of preimplantation genetic testing without success.

Recommendation 5: Before any attempt to establish a pregnancy with an embryo that has undergone genome editing, preclinical evidence must demonstrate that heritable human genome editing (HHGE) can be performed with sufficiently high efficiency and precision to be clinically useful. For any initial uses of HHGE, preclinical evidence of safety and efficacy should be based on the study of a significant cohort of edited human embryos and should demonstrate that the process has the ability to generate and select, with high accuracy, suitable numbers of embryos that:
- have the intended edit(s) and no other modification at the target(s);
- lack additional variants introduced by the editing process at off-target sites—that is, the total number of new genomic variants should not differ significantly from that found in comparable unedited embryos;
- lack evidence of mosaicism introduced by the editing process;
- are of suitable clinical grade to establish a pregnancy; and
- have aneuploidy rates no higher than expected based on standard assisted reproductive technology procedures.

Recommendation 6: Any proposal for initial clinical use of heritable human genome editing should meet the criteria for preclinical evidence set forth in Recommendation 5. A proposal for clinical use should also include plans to evaluate human embryos prior to transfer using:
- developmental milestones until the blastocyst stage comparable with standard in vitro fertilization practices; and
- a biopsy at the blastocyst stage that demonstrates
  o the existence of the intended edit in all biopsied cells and no evidence of unintended edits at the target locus; and
  o no evidence of additional variants introduced by the editing process at off-target sites.

If, after rigorous evaluation, a regulatory approval for embryo transfer is granted, monitoring during a resulting pregnancy and long-term follow-up of resulting children and adults is vital.

Recommendation 7: Research should continue into the development of methods to produce functional human gametes from cultured stem cells. The ability to generate large numbers of such stem cell–derived gametes would provide a further option for prospective parents to avoid the inheritance of disease through the efficient production, testing, and selection of embryos without the disease-causing genotype. However, the use of such in vitro–derived gametes in reproductive medicine raises distinct medical, ethical, and societal issues that must be carefully evaluated, and such gametes without genome editing would need to be approved for use in assisted reproductive technology before they could be considered for clinical use of heritable human genome editing.

Recommendation 8: Any country in which the clinical use of heritable human genome editing (HHGE) is being considered should have mechanisms and competent regulatory bodies to ensure that all of the following conditions are met:

- individuals conducting HHGE-related activities, and their oversight bodies, adhere to established principles of human rights, bioethics, and global governance;
- the clinical pathway for HHGE incorporates best practices from related technologies such as mitochondrial replacement techniques, preimplantation genetic testing, and somatic genome editing;
- decision making is informed by findings from independent international assessments of progress in scientific research and the safety and efficacy of HHGE, which indicate that the technologies are advanced to a point that they could be considered for clinical use;
- prospective review of the science and ethics of any application to use HHGE is diligently performed by an appropriate body or process, with decisions made on a case-by-case basis;
- notice of proposed applications of HHGE being considered is provided by an appropriate body;
- details of approved applications (including genetic condition, laboratory procedures, laboratory or clinic where this will be done, and national bodies providing oversight) are made publicly accessible, while protecting family identities;
- detailed procedures and outcomes are published in peer-reviewed journals to provide dissemination of knowledge that will advance the field;
- the norms of responsible scientific conduct by individual investigators and laboratories are enforced;
- researchers and clinicians show leadership by organizing and participating in open international discussions on the coordination and sharing of results of relevant scientific, clinical, ethical, and societal developments impacting the assessment of HHGE's safety, efficacy, long-term monitoring, and societal acceptability;
- practice guidelines, standards, and policies for clinical uses of HHGE are created and adopted prior to offering clinical use of HHGE; and
- reports of deviation from established guidelines are received and reviewed, and sanctions are imposed where appropriate.

Recommendation 9: An International Scientific Advisory Panel (ISAP) should be established with clear roles and responsibilities before any clinical use of heritable human genome editing (HHGE). The ISAP should have a diverse, multidisciplinary membership and should include independent experts who can assess scientific evidence of safety and efficacy of both genome editing and associated assisted reproductive technologies.

The ISAP should:

- provide regular updates on advances in, and the evaluation of, the technologies that HHGE would depend on and recommend further research developments that would be required to reach technical or translational milestones;

continued

BOX S-1 Continued

- assess whether preclinical requirements have been met for any circumstances in which HHGE may be considered for clinical use;
- review data on clinical outcomes from any regulated uses of HHGE and advise on the scientific and clinical risks and potential benefits of possible further applications; and
- provide input and advice on any responsible translational pathway to the international body described in Recommendation 10, as well as at the request of national regulators.

Recommendation 10: In order to proceed with applications of heritable human genome editing (HHGE) that go beyond the translational pathway defined for initial classes of use of HHGE, an international body with appropriate standing and diverse expertise and experience should evaluate and make recommendations concerning any proposed new class of use. This international body should:

- clearly define each proposed new class of use and its limitations;
- enable and convene ongoing transparent discussions on the societal issues surrounding the new class of use;
- make recommendations concerning whether it could be appropriate to cross the threshold of permitting the new class of use; and
- provide a responsible translational pathway for the new class of use.

Recommendation 11: An international mechanism should be established by which concerns about research or conduct of heritable human genome editing that deviates from established guidelines or recommended standards can be received, transmitted to relevant national authorities, and publicly disclosed.

The Connections Between Genetic Changes and Health

The ability to make changes to the human genome with predictable effects on health relies on a detailed understanding of how DNA sequence variation contributes to the occurrence and risk of disease. Monogenic diseases are caused by mutation of one or both copies of a single gene. Examples include muscular dystrophy, beta-thalassemia, cystic fibrosis, and Tay-Sachs disease. With some notable exceptions, monogenic diseases are individually rare, but together the thousands of monogenic diseases impose significant morbidity and mortality on populations. Current knowledge of medical genetics suggests that the possibility of using HHGE to increase the ability of prospective parents to have biologically-related children who will not inherit certain monogenic diseases is a realistic one.

On the other hand, most common diseases are influenced by many common genetic variants that each have a small effect on disease risk. In addition, the risk of developing such diseases is often influenced by environmental factors such as diet and lifestyle choices and by circumstances that are difficult to predict. Editing a gene variant associated with such a polygenic disease will typically have little effect on risk of the disease. Preventing the disease might be expected to require dozens or more different edits, some of which could produce adverse effects because of other biological roles the gene may play and other genetic networks with which it interacts. Scientific knowledge is not at a stage at which HHGE for polygenic diseases can be conducted effectively or safely. Similarly, there is insufficient knowledge to permit consideration of genome editing for other purposes, including nonmedical traits or genetic enhancement, because anticipated benefits in one domain might often be offset by unforeseen impact on risk of other diseases. Moreover, for these latter purposes the barrier to social acceptability would be particularly high.

Undertaking Genome Editing and Characterizing Its Effects

At present, the primary approach that could be used for undertaking HHGE would involve genome editing in zygotes. A zygote is the single, fertilized cell that results from the combination of parental gametes—the egg and sperm—and is the earliest stage in embryonic development. Although the pace of advances in developing genome editing methodologies continues to be rapid, and ongoing research to overcome current scientific and technical challenges will continue to be valuable, significant knowledge gaps remain concerning how to control and characterize genome editing in human zygotes, as well as in the development of potential alternatives to zygote editing. Gaps that would need to be addressed include the following:

Limitations in the Understanding of Genome Editing Technologies. The outcomes of genome editing in human zygotes cannot be adequately controlled. No one has demonstrated that it is possible to reliably prevent (1) the formation of undesired products at the intended target site; (2) the generation of unintentional modifications at off-target sites; and (3) the production of mosaic embryos, in which intended or unintended modifications occur in only a subset of an embryo's cells—the effects of such mosaicism are difficult to predict. An appropriately cautious approach to any initial human uses would include stringent standards for preclinical evidence on each of these points.

Limitations Associated with Characterizing the Effects of Genome Editing in Human Embryos. Protocols suitable for preclinical validation of human editing would need to be developed to determine (1) the efficiency of achieving desired on-target edits, (2) the frequency with which undesired edits are made, and (3) the frequency with which mosaic editing occurs.

Recommendation 1: No attempt to establish a pregnancy with a human embryo that has undergone genome editing should proceed unless and until it has been clearly established that it is possible to efficiently and reliably make precise genomic changes without undesired changes in human embryos. These criteria have not yet been met, and further research and review would be necessary to meet them.

IMPORTANCE OF SOCIETAL DECISION MAKING ABOUT HERITABLE HUMAN GENOME EDITING

This report focuses on whether a responsible translational pathway can be defined for some potential applications of HHGE. However, it is important to emphasize that the existence of a responsible clinical translational pathway does not mean that a clinical use of HHGE should proceed. Before any such clinical use, there must be widespread societal engagement and approval, and the establishment of national and international frameworks for responsible uses. This Commission highlights the importance of these societal considerations, while acknowledging that the appropriate mechanisms for addressing them lie beyond its charge.

Recommendation 2: Extensive societal dialogue should be undertaken before a country makes a decision on whether to permit clinical use of heritable human genome editing (HHGE). The clinical use of HHGE raises not only scientific and medical considerations but also societal and ethical issues that were beyond the Commission's charge.

CATEGORIZING POTENTIAL USES OF HERITABLE HUMAN GENOME EDITING

Prospective parents who know they are at risk of having a child affected by a monogenic disease already have various reproductive options. Among them is the use of in vitro fertilization together with preimplantation genetic testing (PGT) to ensure that embryos judged suitable for transfer do not carry the disease genotype. In rare cases, every embryo a couple can produce will inherit the disease-causing genotype; for such prospective parents, HHGE could represent the only option to have a genetically-related child without the disease.

In all other groups of prospective parents, some of the embryos are expected not to carry the disease genotype, so PGT can enable them to have an unaffected child. However, a combination of genetic circumstances and reduced fertility can mean that PGT does not always result in the identification of an unaffected embryo for transfer. If HHGE could be performed safely, accurately, and without damaging embryos, it might be possible to increase the number of embryos without a disease genotype that could be used to establish a pregnancy, thereby decreasing the number of treatment cycles required. Whether a meaningful increase could be achieved is currently unclear and would need to be established empirically.

It is not possible to perform a generic benefit–harm analysis covering all possible applications of HHGE since any assessment will depend on the particular circumstances under consideration. One overarching principle that guided the Commission in identifying circumstances for which a responsible translational pathway could be defined was that the highest priority should be given to safety, with any initial uses offering the most favorable balance of potential harms and benefits.

> **Recommendation 3:** It is not possible to define a responsible translational pathway applicable across all possible uses of heritable human genome editing (HHGE) because the uses, circumstances, and considerations differ widely, as do the advances in fundamental knowledge that would be needed before different types of uses could be considered feasible.
>
> Clinical use of HHGE should proceed incrementally. At all times, there should be clear thresholds on permitted uses, based on whether a responsible translational pathway can be and has been clearly defined for evaluating the safety and efficacy of the use, and whether a country has decided to permit the use.

> **Recommendation 4:** Initial uses of heritable human genome editing (HHGE), should a country decide to permit them, should be limited to circumstances that meet all of the following criteria:
> 1. the use of HHGE is limited to serious monogenic diseases; the Commission defines a serious monogenic disease as one that causes severe morbidity or premature death;
> 2. the use of HHGE is limited to changing a pathogenic genetic variant known to be responsible for the serious monogenic disease to a sequence that is common in the relevant population and that is known not to be disease-causing;
> 3. no embryos without the disease-causing genotype will be subjected to the process of genome editing and transfer, to ensure that no individuals resulting from edited embryos were exposed to risks of HHGE without any potential benefit; and
> 4. the use of HHGE is limited to situations in which prospective parents (i) have no option for having a genetically-related child that does not have the serious monogenic disease, because none of their embryos would be genetically unaffected in the absence of genome editing; or (ii) have

extremely poor options, because the expected proportion of unaffected embryos would be unusually low, which the Commission defines as 25 percent or less, and have attempted at least one cycle of preimplantation genetic testing without success.

The report describes six categories of potential uses of HHGE, reflective of these four criteria:

(A) cases in which all of the prospective parents' children would inherit the disease-causing genotype for a serious monogenic disease (defined in this report as a monogenic disease that causes severe morbidity or premature death);
(B) cases in which some but not all of the prospective parents' children would inherit the pathogenic genotype for a serious monogenic disease;
(C) cases involving other monogenic conditions with less serious impact;
(D) cases involving polygenic diseases;
(E) cases involving other applications of HHGE, including changes that would enhance or introduce new traits or attempt to eliminate certain diseases from the human population; and
(F) the special circumstance of monogenic conditions that cause infertility.

To meet all four criteria in Recommendation 4, and based on the available information, the Commission concluded that it is possible to define a responsible translational pathway for initial uses only in Category A and a very small set of circumstances in Category B. To meet the criteria in Category B, reliable methods would need to be developed to ensure that no individuals resulted from embryos that had been subjected to potential adverse consequences of genome editing without potential benefit. Such methods would depend either on identifying zygotes or embryos with the disease-causing genotype before performing HHGE or on excluding from transfer embryos that had needlessly undergone editing.

The Commission concluded that it was not currently possible to define a responsible translational pathway for initial clinical uses of HHGE for other circumstances.

A TRANSLATIONAL PATHWAY FOR HERITABLE HUMAN GENOME EDITING

By a translational pathway for HHGE, the Commission means the steps that would be needed to enable a proposed clinical use to proceed from preclinical research to application in humans. The framework proposed by the Commission draws on experiences of developing a translational pathway for mitochondrial replacement techniques, other ARTs, and from prior clinical experience in editing human somatic cells. If deemed acceptable by a country, HHGE would entail a form of ART used to generate and transfer to the uterus an embryo with an altered genome, resulting in the birth of an individual with this altered DNA.

A translational pathway for uses of HHGE would involve multiple stages (see Figure S-1). Preclinical evidence would need to be obtained from laboratory studies in cultured cells, editing in non-germline human tissues, studies in animal models, and laboratory research in early human embryos. These studies would need to establish that the desired edits can be made reliably, without additional alterations to the genome, and that the process does not alter normal development.

Should a country permit the clinical evaluation of HHGE and should relevant national regulatory authorities give authorization for initial human uses, an embryo with an edited genome would be created with the aim of transferring it to establish a pregnancy. Clinical testing would be undertaken to verify that the embryo had the desired genetic edit and no detectable additional changes that could cause potential harm. Other essential components of any pathway, such as plans for obtaining informed consent and for undertaking short-term and long-term follow-up, would also be evaluated by the regulatory authority as part of the clinical approval process.

SCIENTIFIC VALIDATION AND STANDARDS FOR ANY PROPOSED USE OF HERITABLE HUMAN GENOME EDITING

The initial use of HHGE would represent a new technological intervention in the ART clinic, with only preclinical data with which to judge efficacy and safety. The goal of setting technical standards for HHGE would be to provide very high confidence that any transferred embryos would be correctly edited and that these embryos would have no additional potentially harmful changes introduced by the editing process. For any initial human uses, the standards would need to be set very high, because safety and efficacy could only be fully determined through human use. Preclinical and clinical research must be performed in accordance with the requirements of Recommendation 8.

Recommendation 5: Before any attempt to establish a pregnancy with an embryo that has undergone genome editing, preclinical evidence must demonstrate that heritable human genome editing (HHGE) can be performed with sufficiently high efficiency and precision to be clinically useful. For any initial uses of HHGE,

FIGURE S-1 The main elements of a clinical translational pathway for a proposed use of HHGE to enable parents to have a genetically-related child without a serious monogenic disease. The Commission's work focused on the clinical pathway elements on the right side.

preclinical evidence of safety and efficacy should be based on the study of a significant cohort of edited human embryos and should demonstrate that the process has the ability to generate and select, with high accuracy, suitable numbers of embryos that:

- have the intended edit(s) and no other modification at the target(s);
- lack additional variants introduced by the editing process at off-target sites—that is, the total number of new genomic variants should not differ significantly from that found in comparable unedited embryos;
- lack evidence of mosaicism introduced by the editing process;
- are of suitable clinical grade to establish a pregnancy; and
- have aneuploidy rates no higher than expected based on standard assisted reproductive technology procedures.

Recommendation 6: Any proposal for initial clinical use of heritable human genome editing should meet the criteria for preclinical evidence set forth in Recommendation 5. A proposal for clinical use should also include plans to evaluate human embryos prior to transfer using:

- developmental milestones until the blastocyst stage comparable with standard in vitro fertilization practices; and
- a biopsy at the blastocyst stage that demonstrates
  o the existence of the intended edit in all biopsied cells and no evidence of unintended edits at the target locus; and
  o no evidence of additional variants introduced by the editing process at off-target sites.

If, after rigorous evaluation, a regulatory approval for embryo transfer is granted, monitoring during a resulting pregnancy and long-term follow-up of resulting children and adults is vital.

FUTURE DEVELOPMENTS AFFECTING REPRODUCTIVE OPTIONS

Genome editing in precursor cells that can form eggs and sperm or editing of pluripotent stem cells followed by differentiation into functional gametes *in vitro* (in vitro–derived gametogenesis, IVG) represent potential alternatives to zygote genome editing for HHGE. The technologies to develop human gametes from cultured cells are still under development and are currently unavailable for clinical use. The same is true for the theoretical possibility of extracting human spermatogonial stem cells (SSCs), performing genome editing on them, and reimplanting them in the testes. Any future clinical use of IVG or reimplanted SSCs raises scientific and ethical issues that would require careful consideration, and the procedure would require approval as an ART before it could be used for HHGE.

Genome editing using IVG could address many technical challenges associated with genome editing in zygotes. Methods for characterizing on- and off-target editing are well documented in cultured cells, and only correctly edited cells could be selected and differentiated into functional gametes. Mosaicism would not be an issue when a single sperm derived from an edited induced pluripotent stem cell (iPSC) is used to fertilize a single egg. However, iPSCs and gametes produced from them are likely to undergo adaptation to and expansion in cell culture, which may introduce other types of genetic and epigenetic changes that would need to be carefully assessed.

> **Recommendation 7**: Research should continue into the development of methods to produce functional human gametes from cultured stem cells. The ability to generate large numbers of such stem cell–derived gametes would provide a further option for prospective parents to avoid the inheritance of disease through the efficient production, testing, and selection of embryos without the disease-causing genotype. However, the use of such in vitro–derived gametes in reproductive medicine raises distinct medical, ethical, and societal issues that must be carefully evaluated, and such gametes without genome editing would need to be approved for use in assisted reproductive technology before they could be considered for clinical use of heritable human genome editing.

ESSENTIAL ELEMENTS OF OVERSIGHT SYSTEMS FOR HERITABLE HUMAN GENOME EDITING

From a scientific perspective on safety and efficacy, considerations for any clinical use of HHGE should proceed incrementally. The initial focus would be on potential uses for which available knowledge has established an evidence base that, along with adherence to clinical and ethical norms, makes it possible to define a responsible translational pathway. However, any responsible translational pathway toward potential clinical uses of HHGE requires more than the technical and clinical pathway components. A translational pathway also requires having a comprehensive system for governing any continued development and use of HHGE. It will be important for national and international discussions to establish these governance processes prior to any clinical use under any envisioned circumstance. The work of the WHO's Expert Advisory Committee on Human Genome Editing will be important in this respect.

Governance of HHGE requires a multilayered system of responsibilities. Each country that considers the development of HHGE will end up drawing on the regulatory infrastructure and oversight authorities available under its laws and regulations. But all countries in which HHGE is being researched or conducted would need to have mechanisms in place to oversee translational progress toward potential clinical use of HHGE, to prevent unapproved uses, and to sanction any misconduct. It is recognized that not all countries necessarily have the scientific expertise and regulatory and societal engagement mechanisms to meet the requirements listed below. Nonetheless, if a country is not able to meet all these conditions, no clinical use of HHGE should occur in that country.

> **Recommendation 8:** Any country in which the clinical use of heritable human genome editing (HHGE) is being considered should have mechanisms and competent regulatory bodies to ensure that all of the following conditions are met:
> - individuals conducting HHGE-related activities, and their oversight bodies, adhere to established principles of human rights, bioethics, and global governance;
> - the clinical pathway for HHGE incorporates best practices from related technologies such as mitochondrial replacement techniques, preimplantation genetic testing, and somatic genome editing;
> - decision making is informed by findings from independent international assessments of progress in scientific research and the safety and efficacy of HHGE, which indicate that the technologies are advanced to a point that they could be considered for clinical use;
> - prospective review of the science and ethics of any application to use HHGE is diligently performed by an appropriate body or process, with decisions made on a case-by-case basis;
> - notice of proposed applications of HHGE being considered is provided by an appropriate body;
> - details of approved applications (including genetic condition, laboratory procedures, laboratory or clinic where this will be done, and national bodies providing oversight) are made publicly accessible, while protecting family identities;
> - detailed procedures and outcomes are published in peer-reviewed journals to provide dissemination of knowledge that will advance the field;
> - the norms of responsible scientific conduct by individual investigators and laboratories are enforced;
> - researchers and clinicians show leadership by organizing and participating in open international discussions on the

coordination and sharing of results of relevant scientific, clinical, ethical, and societal developments impacting the assessment of HHGE's safety, efficacy, long-term monitoring, and societal acceptability;

- practice guidelines, standards, and policies for clinical uses of HHGE are created and adopted prior to offering clinical use of HHGE; and
- reports of deviation from established guidelines are received and reviewed, and sanctions are imposed where appropriate.

National decision making should be informed by transparent international discussions before any country's regulatory authorities make major threshold decisions on uses of HHGE. The scientific assessment of whether the suite of technologies on which HHGE would depend have met clear scientific and safety thresholds to be considered for clinical use in a particular set of circumstances will be an essential contribution to both national and international discussions. There is, therefore, a need to regularly review the latest scientific evidence and to evaluate its potential impact on the feasibility of HHGE. The necessary functions of such scientific review include the following:

- assessing or making recommendations on further research developments that would be required to reach technical or translational milestones as research on HHGE progresses;
- providing information to national regulatory authorities or their equivalents to inform their own assessment and oversight efforts;
- facilitating coordination or standardization of study designs to promote the ability to compare and pool data across studies and trans-nationally;
- advising on specific measures to be used as part of the long-term follow-up of any children born following HHGE; and
- reviewing data on clinical outcomes from any regulated uses of HHGE and advising on the potential risks and benefits of possible further applications.

Although there are existing international scientific review bodies that fulfill some of these functions, the Commission does not believe there is an existing mechanism that adequately fulfills all of the functions. The Commission therefore recommends the establishment of a new body, which it has called the International Scientific Advisory Panel.

Recommendation 9: An International Scientific Advisory Panel (ISAP) should be established with clear roles and responsibilities before any clinical use of heritable human genome editing (HHGE).

The ISAP should have a diverse, multidisciplinary membership and should include independent experts who can assess scientific evidence of safety and efficacy of both genome editing and associated assisted reproductive technologies. The ISAP should:

- provide regular updates on advances in, and the evaluation of, the technologies that HHGE would depend on and recommend further research developments that would be required to reach technical or translational milestones;
- assess whether preclinical requirements have been met for any circumstances in which HHGE may be considered for clinical use;
- review data on clinical outcomes from any regulated uses of HHGE and advise on the scientific and clinical risks and potential benefits of possible further applications; and
- provide input and advice on any responsible translational pathway to the international body described in Recommendation 10, as well as at the request of national regulators.

Before crossing any threshold to a new class of use of HHGE, it will be important for the global community to assess not only progress in scientific research but also what additional ethical and societal concerns the circumstances of particular uses could raise, as well as any results, successes, or concerns that had been observed from any human uses of HHGE that had been conducted thus far. New classes of use may or may not precisely align with the six categories defined above. A credible process would need to assess whether it is feasible to envision new translational pathways and what they should entail, and such a body would need not only experts in science, medicine, and ethics but also representatives from the many additional stakeholder communities that could be affected by future uses of HHGE.

Recommendation 10: In order to proceed with applications of heritable human genome editing (HHGE) that go beyond the translational pathway defined for initial classes of use of HHGE, an international body with appropriate standing and diverse expertise and experience should evaluate and make recommendations concerning any proposed new class of use. This international body should:

- clearly define each proposed new class of use and its limitations;
- enable and convene ongoing transparent discussions on the societal issues surrounding the new class of use;

- make recommendations concerning whether it could be appropriate to cross the threshold of permitting the new class of use; and
- provide a responsible translational pathway for the new class of use.

Finally, one other required component of any oversight system is a mechanism for raising concerns about research or clinical use of HHGE, and particularly one allowing a researcher or clinician to bring forward concerns arising from work conducted either in their own or in another country.

Recommendation 11: An international mechanism should be established by which concerns about research or conduct of heritable human genome editing that deviates from established guidelines or recommended standards can be received, transmitted to relevant national authorities, and publicly disclosed.

Introduction

The development of genome editing technologies that have the potential to precisely and efficiently make modifications to DNA inside cells has resulted in renewed attention to the implications of such advances for clinical use in humans. These technologies can be used in different ways; this report focuses on one type of use—making changes to human DNA that could be inherited by future generations. This possibility occurs when genome editing results in the alteration of the DNA in gametes (eggs or sperm) or any cells that give rise to gametes, including the single cell zygote resulting from fertilization of an egg by a sperm cell, or cells of an early embryo. When used clinically, changes to the DNA in such cells can be passed on to the next generation—a process referred to in the report as heritable human genome editing (HHGE) (see Box 1-1).

Germline genome editing is already in use in plant and non-human animal species, primarily in a research context. But the use in humans of heritable genome editing raises many critical and potentially contentious issues. The challenge of assessing safety and efficacy is particularly great, since the effects may not be immediately apparent and could affect future generations. Moreover, individuals' ability to access HHGE, as with other medical technologies, would likely be uneven, raising issues of equity and social justice. Decisions about whether or not to make heritable changes in human DNA sequences and, if so, the nature of genetic changes that should or should not be permitted, requires extensive input from across a country. If extensive societal discussions were to result in approval to consider certain clinical applications of HHGE, it would be essential to have an effective

BOX 1-1
Terminology Used in This Report

Many previous discussions of genome editing have used the terms somatic genome editing and germline genome editing to distinguish non-heritable and heritable applications, respectively. Somatic cells include all the cells of the body except for the germline cells—sperm, eggs, and their precursor cells. Eggs and sperm fuse during sexual reproduction to create a zygote, the initial single cell that continues the germline into the next generation. While heritable human genome editing would necessarily involve using editing reagents with germline cells or their precursors, not all such editing is intended to be inherited. For example, germline genome editing would include any preclinical research that involves genome editing in human zygotes, yet the results of that editing are not inherited by the next generation because it is being done only for research purposes. To distinguish between germline genome editing that is done for research purposes and that done for clinical purposes, the report uses the following terms: (1) the phrase "genome editing in human embryos" or equivalent description when such editing is conducted as part of basic and preclinical laboratory research; and (2) the term "heritable human genome editing (HHGE)" to refer to any editing in germline cells that is done in a clinical context, with the intent of transferring any resultant embryos to a woman's uterus for gestation. As it is conceivable that heritable changes could be made by targeting germline cells in the body of an adult or in an embryo during gestation, the Commission's conclusions and recommendations should be considered as equally applicable to any such *in vivo* applications, although such applications are not discussed further in this report.

translational framework for evaluating the safety and efficacy of the genome editing, assessing the balance of benefits and harms for any given treatment, and overseeing and governing its responsible development and use.

INTERNATIONAL DISCUSSIONS OF
HERITABLE HUMAN GENOME EDITING

There is a long history of discussions on ethical and social implications of making heritable changes to the human genome (Evans, 2002; Fletcher, 1971; Frankel and Chapman, 2000; President's Commission, 1982; Stock and Campbell, 2000), and recent developments in genome editing methods have resulted in renewed urgency in these discussions, which are no longer of purely theoretical interest. Following the demonstration that CRISPR-Cas systems[1] can be used to readily edit the genomes of living human cells,

---

[1] CRISPR stands for clustered regularly interspaced short palindromic repeats, and Cas stands for CRISPR-associated protein. See the glossary at the end of the report for the definition of these and other terms used throughout.

multiple members of the scientific community developing the technology, professional scientific societies, academies of sciences and medicine, bioethics scholars and organizations, and many others convened discussions and published statements and reports addressing the implications of genome editing in humans. For example, the International Bioethics Committee of the United Nations Educational, Scientific, and Cultural Organization issued updated guidance to reflect genome editing advances (UNESCO, 2015). The U.S. National Academy of Sciences and National Academy of Medicine, the U.K.'s Royal Society, and the Chinese Academy of Sciences convened an International Summit on Human Gene Editing that drew more than 3,500 in-person and online participants (NASEM, 2015). More than 60 reports have been published from more than 50 countries dealing wholly or in part with HHGE (e.g., ANM, 2016; Bosley et al., 2015; Brokowski, 2018; CEST, 2019; EGE, 2016; FEAM, 2017; Hinxton Group, 2015; ISSCR, 2015; KNAW, 2016; Lanphier et al., 2015; Leopoldina, 2015; NASEM, 2017). Many groups reiterated that any use of HHGE remained premature and should not be undertaken, with some calling for an explicit moratorium or international prohibition on such use, and others emphasized that HHGE should not be attempted unless or until safety and efficacy were better understood and extensive public engagement and social decision making had taken place. Reports similarly noted the need for appropriate national and transnational oversight and governance structures to be developed prior to any clinical use of HHGE (ISSCR, 2016; NCB, 2016).

In 2017, the U.S. National Academies of Sciences, Engineering, and Medicine released a report authored by an international committee that examined both somatic cell and germline genome editing, possible clinical applications of these technologies, potential risks and benefits, and the regulation of human genome editing (NASEM, 2017). As with prior studies, the report emphasized that any clinical use of HHGE would be premature and that extensive public participation should precede any consideration to authorize clinical trials. However, the report went on to say that HHGE might be permissible sometime in the future, after much more research had been done on balancing risks and benefits, and identified 10 criteria for potential future clinical evaluation of the process as part of a robust regulatory framework. Likewise, in a 2018 extension of its earlier report, the Nuffield Council on Bioethics stated that it could "envision circumstances in which heritable genome editing interventions should be permitted" (NCB, 2018, p. 154). However, such uses would need to safeguard the welfare of people affected by such interventions and not produce or exacerbate social divisions or the marginalization of disadvantaged groups within a country.

CLINICAL USE OF
HERITABLE HUMAN GENOME EDITING REPORTED

In 2018, on the eve of the Second International Summit on Human Genome Editing in Hong Kong, a scientist working in Shenzhen, China, announced that he had used genome editing tools to make alterations in early human embryos that were subsequently transferred to the intended mother, resulting in the birth of twin girls (NASEM, 2019b). At the summit, the researcher revealed that a second pregnancy had been established using a similarly-edited embryo. In his presentation at the summit, he described how his research team had "conducted experiments on embryos from mice and monkeys, human embryonic stem cells, and cultured human embryos" to introduce a deletion into the *CCR5* gene, which "plays a role in the infection of cells by human immunodeficiency virus" (NASEM, 2019b, p. 2). Concluding that the procedure was safe, the researcher and his associates used CRISPR-Cas9 in fertilized human eggs in an attempt to edit *CCR5* and protect the resulting children against infection by this virus. The data presented in Hong Kong revealed that the *CCR5* target was fully modified in only one of the embryos, and the scientists' claims have not been independently and publicly verified (Cohen, 2019a; Cyranoski, 2019). Following an investigation by Chinese authorities, it was announced at the end of 2019 that the researcher and his collaborators had been found guilty of having "forged ethical review documents and misled doctors into unknowingly implanting gene-edited embryos into two women" and had received fines and prison sentences (Normile, 2019).

The response to the news of this clinical use of HHGE was immediate and forceful. Despite what many had viewed as general agreement within scientific and clinical communities that it would be premature and irresponsible to undertake HHGE at this time, it had apparently taken place. In its concluding statement, the summit organizing committee described the reported clinical use of HHGE as "deeply disturbing" and criticized the violation of ethical standards and lack of transparency in the development, review, and conduct of the clinical procedures. It went on to state that clinical trials of HHGE could become acceptable in the future if (1) the risks could be evaluated and satisfactorily addressed, and (2) criteria on societal acceptability were met. It suggested that "it is time to define a rigorous, responsible translational pathway toward such trials" (NASEM, 2019b, p. 7).

More than 100 Chinese scientists signed an online declaration calling the work "crazy" (Cohen, 2019b). Scientists declared that such an experiment on human beings is not morally or ethically defensible. There were renewed calls for a global moratorium on clinical use of HHGE for a defined period of time, to allow time to develop international guidelines

(ASGCT, 2019; Lander, et al., 2019). The German Ethics Council echoed the call for an international moratorium and recommended that an international oversight agency be established to develop standards by which such interventions could be administered, should they be determined to be safe, efficacious, and permissible (GEC, 2019). While supporting the need to address HHGE, others argued against declaring a moratorium because of concerns that it would be open-ended in duration, could impede scientific research, and could be less effective than developing stringent oversight systems (Adashi and Cohen, 2019).

FORMATION OF THE INTERNATIONAL COMMISSION AND WORLD HEALTH ORGANIZATION EXPERT COMMITTEE

Two international committees were convened following these calls to further develop an understanding of what would be involved in a responsible translational pathway toward HHGE and to make progress on effective coordination and governance of human genome editing.

The International Commission on the Clinical Use of Human Germline Genome Editing (the Commission authoring the present report) was convened by the U.S. National Academy of Medicine, the U.S. National Academy of Sciences, and the U.K.'s Royal Society. The Commission has been tasked with developing a framework for scientists, clinicians, and regulatory authorities to consider when assessing potential clinical applications of HHGE. This framework could be used in the development of a potential pathway from research to clinical use, should a country conclude that HHGE applications are acceptable. The Commission's goal is to prepare the way for international agreement on specific criteria and standards that would have to be met before HHGE could be deemed permissible, if permissible at all.

The World Health Organization (WHO) also established a global multidisciplinary expert committee to examine the scientific, ethical, social, and legal challenges associated with human genome editing—both somatic and heritable (WHO, 2019b). The Expert Advisory Committee on Human Genome Editing will advise the director general of the WHO on appropriate oversight and governance mechanisms, both at the national and global levels.

While the deliberations of the Academies' International Commission and the WHO's Expert Advisory Committee are likely to overlap to some extent with respect to HHGE, the WHO Committee's focus is on governance mechanisms, while the Academies' Commission is more concerned with the scientific and technological questions that would need to be addressed as part of such governance. The WHO Committee will also consider the broader social and ethical questions raised by the possible use of

HHGE, whereas this Commission's mandate is limited to issues inextricably linked to research and clinical practice.

This report has been released while the WHO Committee is still developing its recommendations and is intended to inform that committee's deliberations. It should also be relevant to national and international policy makers as they consider laws and regulatory frameworks for HHGE. Inevitably, it provides a current snapshot of the relevant technologies and addresses only some of the issues that policy makers will need to take into account.

MITOCHONDRIAL REPLACEMENT TECHNIQUES: MODIFYING THE EMBRYO

In developing its recommendations, the Commission sought to learn from prior experience with related technologies. Mitochondrial replacement techniques (MRT) constitute the only technology currently approved anywhere in the world that results in genetic changes that can be inherited. The approval of MRT for clinical use in the United Kingdom was driven by patient need and was introduced only after extensive preclinical research and consideration by a regulatory body already established for the oversight of assisted reproductive technologies (ARTs).[2] The principle of MRT is illustrated in Figure 1-1.

In addition to the DNA contained in chromosomes in the cell nucleus (the nuclear genome), eukaryotic cells also contain hundreds or thousands of DNA molecules in organelles called mitochondria; these DNA molecules constitute the mitochondrial genome. People inherit their mitochondrial DNA (mtDNA) only from their biological mother, since sperm mitochondria are eliminated during embryo development. Disease-causing mtDNA mutations can occur in all or a fraction of the mitochondria in a person's cells.[3] These mutations can cause a wide variety of human diseases for which little or no treatment is currently available. One in every 5,000 to 10,000 people develops a symptomatic mtDNA disease.

In one method of MRT, referred to as maternal spindle transfer, eggs are harvested from the intended mother who has pathogenic mtDNA, and the chromosomes of the nuclear genome of each egg are removed and transferred to the donated eggs of a woman with healthy mtDNA from which the chromosomes of the nuclear genome have been removed. After fertil-

---

[2] Additional information is available at https://www.hfea.gov.uk/treatments/embryo-testing-and-treatments-for-disease/mitochondrial-donation-treatment/.

[3] A change in DNA sequence that produces a change in phenotype is referred to in this report as a "mutation" or a "pathogenic variant." Other changes in DNA sequence that are typically common in the population and have little or no effect on disease risk are referred to as "variants."

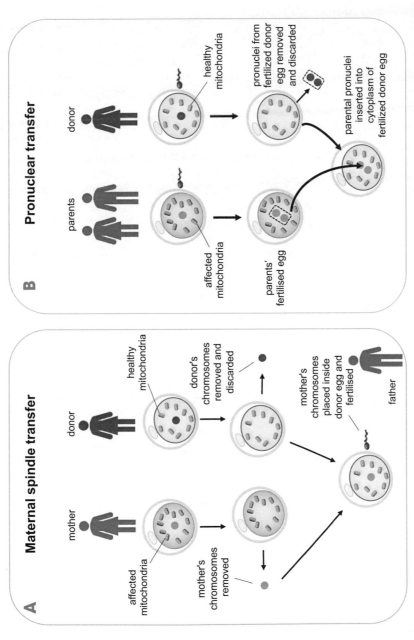

FIGURE 1-1 Methods for MRT include (A) maternal spindle transfer and (B) pronuclear transfer. For details, see Greenfield et al. (2017).

ization, the resulting embryos are transferred to the intended mother. An alternative method of MRT, called pronuclear transfer, involves transferring chromosomes of the nuclear genome between zygotes rather than between unfertilized eggs. With both methods, a child born using MRT has nuclear DNA from the child's mother and father and mtDNA from the egg donor.

With MRT, no DNA sequences are directly altered; rather, entire chromosomes are transferred from one egg to another. By contrast, HHGE makes it possible to alter the DNA sequence of any of the 6 billion base pairs comprising the complete set of the mother's and father's chromosomes.

MRT was first legally approved for clinical use in the United Kingdom and is carried out under the regulatory framework of the U.K. Human Fertilisation and Embryology Authority (HFEA), which grants approval for treatment on a case-by-case basis. MRT is only permitted for the prevention of serious mtDNA disease, with the additional caveats that licenses can only be granted to named clinics after demonstration of their competence, and any use is restricted to prospective parents with no suitable alternative for having an unaffected, genetically-related child. The steps involved in developing a translational pathway for MRT in the United Kingdom are summarized in Box 1-2. Despite the important differences between MRT and HHGE, this pathway can provide insights to inform similar debates about HHGE.

BOX 1-2
The Pathway Toward the Regulated Use of Mitochondrial Replacement Techniques in the United Kingdom

Developing a translational pathway for the use of MRT involved important elements over multiple years. These included the following:

- **A legal and regulatory foundation.** All human embryo research and the use of ARTs are subject to the Human Fertilisation and Embryology Act. The Act became law in 1990 and enabled creation of HFEA as the statutory regulator. The Act and its subsequent amendments set out the licensing regime for producing an embryo using ARTs for research or to start a pregnancy. Licenses can only be granted for a set of approved purposes and procedures and to a named institution where the procedures are carried out under the guidance of a Person Responsible. Without such licenses, all uses of human embryos to carry out research or establish a pregnancy are legally prohibited.

continued

BOX 1-2 Continued

- **Initial demonstration of potential feasibility.** The U.K. chief medical officer's report *Stem Cell Research: Medical Progress with Responsibility* (UKDH, 2000) discussed the possibility of MRT to prevent the inheritance of diseases caused by mutations in mtDNA, and MRT had previously been demonstrated in animal models. In 2005, Newcastle University received the first HFEA license to carry out research demonstrating the feasibility of MRT using human embryos.
- **Support from patient communities seeking MRT to prevent disease and address unmet clinical needs.** Moves toward permitting MRT were largely driven by a patient community, people unable to have healthy, genetically-related children due to mtDNA disease.
- **Public engagement and ethical dialogues.** In 2012, the U.K. Government asked the HFEA to commission a public dialogue to explore views on the possible use of MRT. In parallel, the U.K.'s Nuffield Council on Bioethics conducted an inquiry into ethical issues raised by MRT. Based on the public dialogue and Nuffield Council report, the HFEA advised the Government that "there is general support for permitting mitochondrial replacement in the U.K., so long as it is safe enough to offer in a treatment setting and is done so within a regulatory framework" (HFEA, 2013, p. 4).
- **Legislative approvals.** In 2008, the Human Fertilisation and Embryology Act was amended to enable clinical use of MRT, should the U.K. Parliament approve it. In 2014, the Government drafted regulations that would enable the clinical use of MRT. These went through the legislative process, including a period of public consultation and a series of parliamentary debates, and were passed in 2015. The regulations set out the specific technologies that could be used to prevent the transmission of serious mtDNA disease and ensured that the HFEA had oversight of this novel ART.
- **Independent expert reviews of safety and efficacy.** The HFEA commissioned four independent expert reviews of the science and technology of MRT (HFEA, 2011, 2013, 2014, 2016). These examined the techniques available for carrying out MRT, whether they were likely to be effective at preventing the inheritance of mitochondrial diseases, and whether the processes themselves could lead to harm. The expert review in 2016 concluded that the techniques were safe enough for limited clinical use.
- **Regulatory review and approval for clinical use on a case-by-case basis.** The first license to carry out MRT was awarded by the HFEA in 2017 to the Newcastle Fertility Centre. Since then, 20 approvals for individual couples to be treated have been granted, but no further details are available due to issues of confidentiality. A condition of the HFEA license granted to the centre is that a clinical pathway be in place to ensure that all pregnancies are carefully monitored and that a process be in place for the long-term follow-up of individuals born.

A TRANSLATIONAL PATHWAY FOR
HERITABLE HUMAN GENOME EDITING

By a "translational pathway" for HHGE, the Commission means the steps that would be needed to enable a proposed clinical use to proceed from preclinical research to application in humans. Elements that formed the pathway leading to clinical use of MRT in the United Kingdom have informed the Commission's development of a clinical pathway toward HHGE, presented in this report. The core elements of this pathway are shown in Figure 1-2 and described below.

Societal Considerations

As demonstrated by the example of MRT, progress through a clinical translational pathway for controversial technologies such as HHGE requires widespread public discussion about whether a technology is broadly acceptable and, if so, for what purposes, with which checks and balances, and with what oversight. These considerations are shown in the orange boxes in Figure 1-2.

The upper box on the left side of the figure represents critical deliberations that would be required prior to any society determining that it would permit the clinical use of HHGE. Such discussions will need to include broad public engagement on the potential uses and implications of HHGE, as well as development of those legislative, regulatory, and institutional foundations that would need to be in place prior to any clinical use.

The lower box reflects the fact that, even were an initial clinical use and evaluation of HHGE to be permitted and undertaken, societal deliberations would need to continue. The outcomes of any initial human use of HHGE would need to be considered, lessons taken into account, and further extensive scientific, clinical, stakeholder, and public input incorporated to decide whether to consider any further clinical uses.

These discussions are as important as the clinical pathway components (on the right side of the figure) but are beyond this Commission's Statement of Task (see Box 1-3). Questions that deserve significant attention include how to effectively engage multiple sectors of the public, including genetic disease and disability communities, and how to incorporate the diverse input received into a country's decision-making processes. Through presentations to the Commission and responses to the Commission's call for evidence, respondents from civil society, including genetic disease and disability communities, shared elements they felt are important to consider. Though addressing these elements was not in the Commission's charge, two themes may inform future HHGE deliberations:

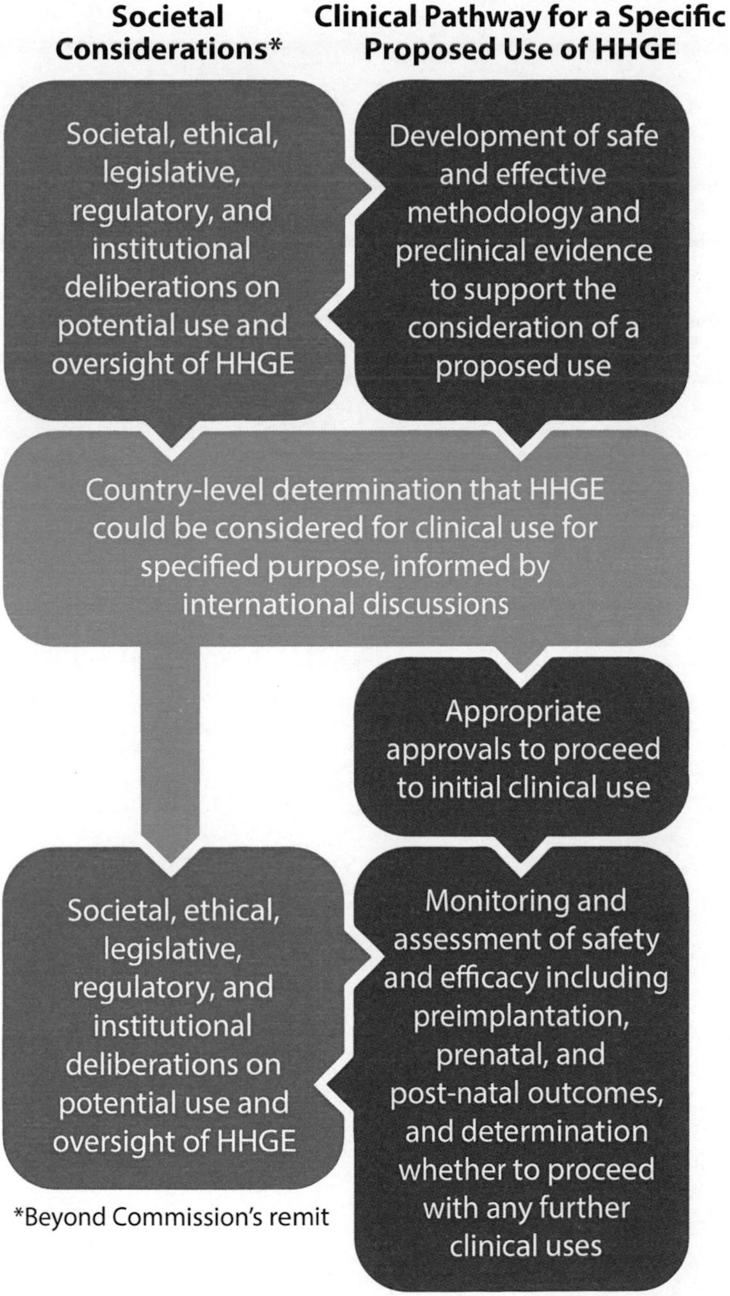

**Societal
Considerations***

**Clinical Pathway for a Specific
Proposed Use of HHGE**

Societal, ethical,
legislative,
regulatory, and
institutional
deliberations on
potential use and
oversight of HHGE

Development of safe
and effective
methodology and
preclinical evidence
to support the
consideration of a
proposed use

Country-level determination that HHGE
could be considered for clinical use for
specified purpose, informed by
international discussions

Appropriate
approvals to proceed
to initial clinical use

Societal, ethical,
legislative,
regulatory, and
institutional
deliberations on
potential use and
oversight of HHGE

Monitoring and
assessment of safety
and efficacy including
preimplantation,
prenatal, and
post-natal outcomes,
and determination
whether to proceed
with any further
clinical uses

*Beyond Commission's remit

FIGURE 1-2 General elements that form a translational pathway for HHGE.

BOX 1-3
Statement of Task

Clinical applications of germline genome editing are now possible, and there is an urgent need to examine the potential of this new technology. Many scientific and medical questions about the procedures remain to be answered, and determining the safety and efficacy of germline genome editing will be necessary but not sufficient conditions for future clinical usage. There is a need for a framework to inform the development of a potential pathway from research to clinical use, recognizing that components of this framework may need to be periodically revised in response to our rapidly evolving knowledge. In addition, other important discussions are ongoing internationally about the implications for society of human germline genome editing and include issues such as access, equity, and consistency with religious views.

An international Commission will be convened with the participation of National Academies of Sciences and Medicine throughout the world to develop a framework for considering technical, scientific, medical, regulatory, and ethical requirements for germline genome editing, should society conclude such applications are acceptable.

The U.S. National Academies of Sciences and Medicine and the U.K. Royal Society will serve as the Commission's secretariat. Specifically, the Commission will:

1. Identify the scientific issues (as well as societal and ethical issues, where inextricably linked to research and clinical practice) that must be evaluated for various classes of possible applications. Potential applications considered should range from genetic correction of severe, highly penetrant monogenic diseases to various forms of genetic enhancement.
2. Identify appropriate protocols and preclinical validation for assessing and evaluating on-target and off-target events and any potential developmental and long-term side effects.
3. Identify appropriate protocols for assessing and evaluating potential mosaicism and long-term implications.
4. Identify ways to assess the balance between potential benefits and harms to a child produced by genome editing and to subsequent generations.
5. Design appropriate protocols for obtaining consent from patients, for obtaining ethical approval from knowledgeable review committees, and for satisfying regulatory authorities.
6. Identify and assess possible mechanisms for the long-term monitoring of children born with edited genomes.
7. Outline the research and clinical characteristics developed in tasks 1–6 that would form part of an oversight structure, including defining scientific criteria for establishing where heritable genome editing might be appropriate, overseeing any human clinical use, and bringing forward concerns about human experiments.

1. *The need for discourse within civil society about human genome editing.* It will be important not to limit the focus of public engagement and civil society discussions to only scientific and clinical dimensions. Discussions must also occur concerning the implications of HHGE for inequity and social justice, the value placed on genetic relatedness of a child and on parental reproductive preferences, societal attitudes toward disease and disease prevention, privacy considerations, religious scholarship, and ethics.
2. *The importance of engaging directly with people who have conditions that might be considered for HHGE.* People who are living with genetic disease or disability must be able to meaningfully participate in societal discussions of genome editing and take part in policy development processes. It is important that not all engagement be led by scientific and clinical communities but also by those whom the technology would most affect.

Decision by a Country to Permit Consideration of Heritable Human Genome Editing for a Proposed Clinical Use

A country's lawmakers need to assess information on the safety and potential therapeutic efficacy of a technology and give consideration to public opinion to decide whether a new medical technology should be made available within its jurisdiction and, if so, for what uses it should be permitted. These considerations are shown in the green box in Figure 1-2. The outcomes of extensive societal deliberations and sufficient progress in preclinical development of the techniques for HHGE would together feed into a country's decision on whether or not HHGE could be considered for clinical use. If a country's legislative body does not permit the consideration of HHGE for the proposed purpose, the pathway toward clinical use cannot proceed beyond basic laboratory research and preclinical development. HHGE currently remains illegal or otherwise not approved in many countries. It is not specifically regulated in other countries, which would need to consider establishing relevant national regulations.

Clinical Pathway for a Specific Proposed Use of Heritable Human Genome Editing

The focus of this Commission's task is the track shown on the right side of Figure 1-2 in the blue boxes. Any pathway for the use of HHGE starts with a specific proposed use. It consists of four primary elements. The blue box at the top of the figure represents the development of preclinical evidence that demonstrates the feasibility of HHGE for the proposed use.

Such evidence would be obtained from laboratory studies in cultured cells, genome editing in other types of non-germline human tissues, studies in animal models, and research in human embryos that are not used to create a pregnancy. Research into heritable genome editing is currently at this stage of the pathway.

Two critical decision points occur in the middle of the pathway. First, as described above, a country must permit the consideration of HHGE for clinical use (green box). Were a country to permit the relevant national regulatory authority to consider a request to use HHGE for a proposed purpose, a second critical decision point is reached (second blue box). Once the preclinical evidence base has been established and a particular methodology is deemed to meet safety and efficacy thresholds, an application may be made to the appropriate regulatory body requesting the opportunity to use the technology in humans. Any clinical team seeking to proceed with an initial use of HHGE would need to receive scientific and ethics approvals from institutional and/or national advisory bodies or regulatory authorities prior to undertaking any clinical use.

Finally, should such approvals be granted, clinical use of HHGE for the proposed purpose could be undertaken (lower blue box). This is the stage at which a human pregnancy would be attempted by uterine transfer of an embryo whose genome had undergone editing. Further evidence on the safety and efficacy of the technology would be obtained, including through monitoring during the pregnancy and by follow-up of individuals born with edited genomes. Information on the outcomes would be integrated into the overall evidence base, and further deliberations would be needed to decide whether to undertake future clinical uses.

STUDY FOCUS AND APPROACH

The Commission's full Statement of Task is provided in Box 1-3. The Commission is international in its mandate and composition, with membership spanning 10 nations and 4 continents and including experts in science, medicine, genetics, psychology, ethics, regulation, and law. The Commission's deliberations included, among other activities, a public meeting held in August 2019, a request for public input to targeted questions gathered in September 2019, public webinars on genome editing technology held in October 2019, and a public workshop held in November 2019. At a third meeting in January 2020, Commission members developed the findings, conclusions, and recommendations presented in this report. See Appendix A for further information on how the Commission conducted its work and Appendix B for brief biographies of Commission members.

This report cannot serve as a checklist that encompasses the details of every experiment, method, or process that would need to be carried out for genome editing to go from laboratory research to clinical use in human embryos. The science of genome editing is advancing so rapidly that new methods and data are reported weekly. The regulatory environments for both genome editing and ARTs vary widely across the globe, and whether HHGE would ever be permitted or how it would be overseen by a country remains to be determined. It is premature to establish specific "protocols" for many of the tasks identified in Box 1-3. Instead, the report describes the key elements that would need to form the foundation for a potential translational pathway for HHGE, lays out scientific and clinical issues that will need to be considered to undertake HHGE responsibly, and identifies preclinical and clinical requirements that would need to be met to establish safety and efficacy.

ORGANIZATION OF THE REPORT

This chapter introduces what the Commission means by a translational pathway toward possible future clinical uses of HHGE, should a country ever permit such use. The subsequent chapters of the report explore components of the pathway in greater detail. The report discusses the current state of the science and whether sufficiently safe and effective editing methodologies currently exist, along with circumstances associated with various types of proposed uses of HHGE. The report also specifies what preclinical evidence and clinical protocols would be required, and what associated oversight frameworks would be needed for any potential clinical use of HHGE.

Chapter 2 provides an overview of areas of science and technology involved in heritable genome editing. Sections address what is known about the genetics of human diseases, reproductive technologies (including in vitro fertilization and preimplantation genetic testing), and genome editing technologies and methodologies for characterizing their effects, and what is known about the possibilities of human genome editing in both embryos and gametes. It concludes by identifying key knowledge gaps that would need to be filled prior to considering any clinical use.

Chapter 3 classifies potential categories of uses for HHGE, each having distinct characteristics. It discusses scientific and clinical issues that would need to be considered in any assessment of the potential use of HHGE in these categories. Based on the current state of understanding, it concludes by identifying categories of potential uses for which the Commission felt it was possible to describe a responsible clinical translational pathway toward an initial human use of HHGE, should a country conclude that such applications are acceptable.

Chapter 4 describes the preclinical and clinical requirements that would need to be met as part of the potential translational pathway for the initial clinical use of HHGE described in this report.

Finally, Chapter 5 sets out recommendations on requirements for oversight frameworks in any country that is considering enabling the clinical use of HHGE. It also emphasizes the need for international coordination and makes recommendations for core components of such efforts. The establishment of oversight mechanisms and infrastructure to govern the use of HHGE is critical to any responsible translational pathway and for preventing misuse of the technology.

The State of the Science

Chapter 2 provides the foundations in areas of science and medicine that are important for understanding the feasibility of heritable human genome editing (HHGE). This chapter contains substantial scientific detail; see the glossary in Appendix C for any unfamiliar terms. Part I of this chapter describes what is known about the genetics of diseases caused by mutations in a single gene—a category known as monogenic diseases. It then discusses potential reproductive options for parents at risk of passing on a disease genotype, including the use and current limitations of in vitro fertilization (IVF) in conjunction with preimplantation genetic testing (PGT) to identify any embryos that do not have the disease-causing genotype.

Part II reviews genome editing technologies and current approaches to characterizing their results. It describes what has been learned so far from genome editing in somatic cells and in early embryos. Genome editing carried out concomitant with fertilization, or in a zygote (the single cell created by fertilization), would be the most likely way in which the potential for clinical use of HHGE could currently be evaluated.

Part III discusses a technology with the potential to provide another means of preventing the inheritance of genetic disorders, as well as an alternative to zygote genome editing for undertaking HHGE: the ability to create sperm or egg cells in the laboratory from parental stem cells. At this time, further development would be required before this technology could be considered for clinical use. Even then, it would have significant scientific, ethical, and social implications. As with HHGE, the decision about whether to make it available for clinical use would depend on much more than technological feasibility.

Part IV reviews two other areas that would be crucial components of any clinical use of HHGE. Informed consent would be needed from prospective parents, and HHGE would pose special challenges to these protocols. In addition, the long-term monitoring of any individuals born following HHGE would be important, and such monitoring would potentially span generations, raising further issues.

Part V reflects on the complexities of human genetics beyond monogenic diseases and looks ahead to other circumstances for which HHGE has been proposed. It describes what is known about the genetics of polygenic diseases, a category that includes many common diseases in which multiple genetic variants contribute to overall disease risk. And it discusses the genetics of male infertility—a special case for HHGE.

The chapter concludes by identifying key knowledge gaps that would need to be addressed before any clinical use of HHGE and provides two recommendations.

MONOGENIC DISEASES: GENETICS AND REPRODUCTIVE OPTIONS

Genetics of Monogenic Diseases

Over the past 40 years, human genetics has undergone a revolution that has enabled the systematic identification of genes underlying many human diseases (Claussnitzer et al., 2020). The scientific program started with the recombinant DNA revolution in the 1970s, which allowed the cloning and isolation of segments of the genome of any species. This led to recognition that physical and genetic maps of genomes could be unified (Wensink et al., 1974), resulting in the idea of "positional cloning." In this paradigm, the chromosomal location of a trait-causing mutation could be determined by any of several genetic methods, and the cloned DNA segments from the section of the chromosome thought to contain the responsible mutation could then be assembled and analyzed to identify the specific gene and mutation that produce the disease or trait (Bender et al., 1983).

The development of positional cloning in humans became possible with the recognition in 1980 that there is substantial polymorphism in DNA sequence of genomes, with alternative sequences that are common in populations (Botstein et al., 1980). These alternative sequences mark a specific chromosome segment and permit the tracing of its inheritance through pedigrees or populations. The discovery of millions of these common variations allowed the development of genetic maps of the human genome, thereby permitting the systematic comparison of the inheritance of every segment of every chromosome to the inheritance of diseases or other traits in families. For diseases caused by mutation in a single gene, this process demonstrated which chromosome segment was precisely linked with the disease or trait. From this map location, the disease gene could eventually be discovered as the gene in the mapped interval that harbors mutations specific to individuals with the disease or trait. For example, the gene in which mutations lead to cystic fibrosis (CF) was identified in 1989 (Riordan et al., 1989).

This process was greatly accelerated by the assembly of the virtually complete sequence of the human and other genomes, announced in 2001 (IHGSC, 2001, 2004), which greatly aided the discovery of human genes and facilitated the process of identifying disease-causing mutations. These efforts led to the identification of several thousand human genes in which mutation produced a disease phenotype.

Advances in DNA sequencing over the ensuing decade dramatically increased sequence production and reduced its cost by more than a million-fold. This advance led to brute-force methods of disease gene discovery in which sequencing of all ~20,000 protein-coding genes in the human genome in many unrelated patients with the same clinical disease could identify genes that are mutated more often than expected by chance, and also permitted routine establishment of clinical diagnosis of individuals with monogenic diseases.

This work collectively has led to the discovery of the genes responsible for more than 4,000 monogenic (single-gene) diseases to date, such as Duchenne muscular dystrophy, beta-thalassemia, CF, Huntington's disease, and Tay-Sachs disease.[1] As discussed in the last section of this chapter, this work has also advanced our understanding of heart disease and neurodegeneration.

Identifying the genes for monogenic diseases has had profound consequences for medicine. The ability to detect mutations in a gene has enabled clinical diagnostics—for example, early diagnosis available to women with mutations in the gene *BRCA1*, who are at increased risk of breast, ovarian, and other cancers. Biological understanding of disease mechanisms has enabled therapies in some cases, ranging from dietary control (patients with phenylketonuria can avoid severe brain damage by adopting a phenylalanine-restricted diet), to drugs (e.g., the ability to replace missing enzymes, as in Gaucher disease, or to mitigate the impact of mutations that cause CF), to gene-based therapies (e.g., one which delivers to cells a functional copy of a gene missing in spinal muscular atrophy [Hoy, 2019]).

---

[1] See Online Mendelian Inheritance in Man at https://www.omim.org.

The Human Genome

Humans inherit two copies of the genome, one from their mother and one from their father. Each copy of the human genome consists of approximately 3 billion base pairs of genetic information distributed among 23 pairs of chromosomes. Of these, 22 pairs are autosomes (equivalent chromosomes inherited from each parent) and 1 pair comprises the sex chromosomes (X or Y, with females inheriting two X chromosomes and males one X and one Y chromosome). In addition to this nuclear genome, mitochondria in cells contain their own, much smaller genome, as discussed in Chapter 1.

Any two examples of the human genome have around 3 million sequence differences, many of which do not result in observable (phenotypic) effects but which reflect the degree of genetic variation in the human population. The vast majority of these differences are single nucleotide variants (SNVs), in which a single base pair in a specific location in the genome varies among people. Other differences include short insertions or deletions of DNA (indels); longer DNA segments that have been lost, added, duplicated, or transposed; and, at the largest scale, differences in chromosome numbers.

The genetic variation in the human population arises from several factors. Genetic variants originally arise as alterations to the genome sequence that arise during DNA replication or other natural processes. Each individual has an average of about 70 *de novo* SNVs and 6 *de novo* indels not present in their parents (Sasani et al., 2019). The rate of *de novo* mutations is increased in older men due to the high number of cell divisions during spermatogenesis and is referred to as the paternal age effect (reviewed in Cioppi et al., 2019).

Most new variants do not alter reproductive success and are unlikely to persist over time in large populations. For this reason, most common variations found in the human genome were introduced many thousands of years ago, when population sizes were small. Other variants impair reproductive success and are more rapidly eliminated from the population by negative selection. Rarely, a variant will increase reproductive success and thereby increase in frequency in a population over time due to positive selection. Lastly, occasionally a variant will have beneficial effects when it is present in a single copy (allele) but have deleterious effects when present in both alleles, resulting in balancing selection that allows a potentially deleterious variant to be maintained in the population. Over generations, the linkage of these variants to each other on a chromosome is shuffled by genetic recombination between parental chromosome pairs that occurs during the formation of gametes, thereby producing great variation in the combinations of alleles that in turn produce high phenotypic variation in populations.

Monogenic Diseases

Monogenic diseases are caused by mutation of either one or both copies (or alleles) of a single gene, typically by altering the protein-coding sequence of the gene or, less often, by altering a DNA segment that regulates the activity of the gene. The thousands of monogenic diseases vary widely in many respects, including the organ systems affected, the age of onset, and the seriousness of disease.

Some monogenic diseases are caused by dominant mutations. These diseases occur in individuals who carry one disease-causing allele and one non-disease-causing allele in the relevant gene (heterozygotes). An example is Huntington's disease, in which a defect in the gene for a protein active in brain cells gradually causes damage to those cells through the accumulation of the abnormal protein, which leads to progressive neurological symptoms and premature death (Walker, 2007). Other examples include myotonic dystrophy and neurofibromatosis. Dominant diseases can arise because the disease-causing copy of the gene produces too little protein to allow normal function even in the presence of a normal copy of the gene (haploinsufficiency), produces an abnormal protein that interferes with the normal protein produced by the other copy of the gene (dominant negative), or causes too much activity of the normal protein (gain of function), or the abnormal protein acquires a new function, not found in the normal protein, that causes disease (neomorph). In other cases, loss of a single functioning gene copy is tolerated, but the remaining functional copy of the gene is lost in some cells during the lifetime of the individual, leading to disease manifestation restricted to the affected tissue. This is the case in some forms of familial breast and colon cancer.

In other monogenic diseases, the causative mutations are recessive. These diseases occur in individuals who carry disease-causing mutations on both alleles of a gene (mutations are homozygous if the two mutations are identical, or compound heterozygous if they are different). Recessive mutations typically cause loss of normal gene function, as occurs in CF and spinal muscular atrophy, but there are exceptions, such as sickle cell disease (SCD), in which the mutant protein acquires a deleterious function not found in the normal protein.

Still other monogenic diseases are X-linked, due to a mutation in a gene found only on the X chromosome. Males are affected if they carry a mutated allele on their single X chromosome, and females are affected if they carry a disease-causing allele on both of their X chromosomes. Some females who are carriers of the mutated allele may show signs or symptoms of the disease if there is skewed inactivation of their X chromosomes with preferential inactivation of the X chromosome without the mutation (reviewed by Migeon, 2020). Examples include fragile X syndrome, hemophilia A, and Duchenne muscular dystrophy.

In some cases, complexities may be layered over the descriptions above. Monogenic diseases may have incomplete penetrance: only a subset of people who inherit the same disease genotype will actually have the disease. These diseases may also have variable expressivity, and people who inherit the same disease genotype may have different qualitative or quantitative manifestations of the disease. Incomplete penetrance and variable expressivity may be due to the effect of modifier genes elsewhere in the genome, only some of which have been identified. For example, the severity of SCD, caused by mutations in the gene encoding the beta chain of hemoglobin, is modified by genetic variants that affect adult expression of the gene encoding fetal hemoglobin. Disease penetrance and expressivity may also be influenced by non-genetic factors. Well-known examples include phenylketonuria in which an inherited inability to metabolize the amino acid phenylalanine can result in intellectual disability and seizures; however, the disease can be mitigated by a diet low in phenylalanine. Similarly, some immune deficiencies may have no significant clinical consequence unless an individual is exposed to a particular infectious agent such as tuberculosis or influenza.

A single gene can also have different pathogenic variants, some that are more common in particular populations and some that are rare or unique to one or a small number of families. In general, for a gene whose mutation causes a recessive disease, many different disease-causing mutations will be found in populations because there are many ways to produce loss of function mutations in a gene: these can be produced by different premature termination, splice site or frameshift mutation at many different sites along the gene, and by many different protein-altering mutations. The high diversity of these mutations may complicate editing efforts since the required editing reagents for the same gene in different cases could often be different. The same applies to dominant mutations caused by haploinsufficiency. In contrast, dominant diseases caused by gain of function mutations typically have a more restricted spectrum of disease-causing mutations because markedly increasing the activity or producing a distinct function of an encoded protein by mutation is genetically much less frequent than simply knocking out a gene's function.

Nonetheless, some recessive mutations can dominate the allele spectrum in certain diseases. One example is SCD, in which one copy of the hemoglobin S allele can provide some protection against malaria while two mutant copies cause SCD, featuring severe morbidity and premature death (Archer et al., 2018). In this disease, most affected people in or descendent from West Africa have the same disease-causing mutation in beta-hemoglobin. Another serious red blood cell disease, thalassemia, also has relatively frequent alleles, again owing to protection from malaria. Similarly, while more than 1,500 different loss of function mutations in the *CFTR* gene can cause the recessive disease CF, a specific deletion of three nucleotides in this gene comprises approximately 70 percent of all loss of function mutations in *CFTR* in people of northern and central European descent (European Working Group on Cystic Fibrosis Genetics, 1990), while a different mutation is enriched in people of African ancestry.

Inheritance Patterns of Monogenic Diseases

With some notable exceptions, monogenic diseases are individually very rare—with frequencies typically in the range of 1 in 10,000 to 1 in 1 million births.[2] However, the thousands of rare monogenic diseases together impose a significant burden on human health. According to the World Health Organization (2019a), the global prevalence of all monogenic diseases at birth is about 1 in 100, and monogenic conditions have been reported to "collectively contribute to disease in ~0.4 percent of children and young adults" (Posey et al., 2019). In addition, as noted above, there are circumstances in which a monogenic disease is found at higher frequency in a particular population in which the heterozygous state confers an advantage, where a mutation was present in an individual whose genes were inherited by a significant proportion of a population (also known as a founder effect), or where there are high rates of consanguinity (see Chapter 3 for further discussion of circumstances in which certain monogenic diseases are found at higher frequencies).

The typical situation for the inheritance of autosomal dominant and autosomal recessive diseases is shown in Figure 2-1. For an autosomal dominant disease, if one parent is a heterozygote for the disease-causing allele, each offspring of this parent has a 50 percent chance of inheriting a disease-causing genotype and a 50 percent chance of not inheriting a disease-causing genotype. In rare circumstances in which both parents have the same autosomal dominant disease, each offspring would have a 75 percent chance of inheriting the disease-causing genotype (i.e., at least one disease-causing allele). For an autosomal recessive disease, if both parents are unaffected heterozygous carriers, each offspring would have a 25 percent chance of inheriting the disease-causing genotype (i.e., two disease-causing alleles).

There are very rare circumstances, however, in which *all* of a couple's children would inherit the disease genotype, as shown in Figure 2-2. Specifically, these circumstances involve either one parent being homozygous for a dominant disease or both parents being homozygous or compound heterozygous for the same recessive disease.

---

[2] See Online Mendelian Inheritance in Man (OMIM) at https://www.omim.org.

42

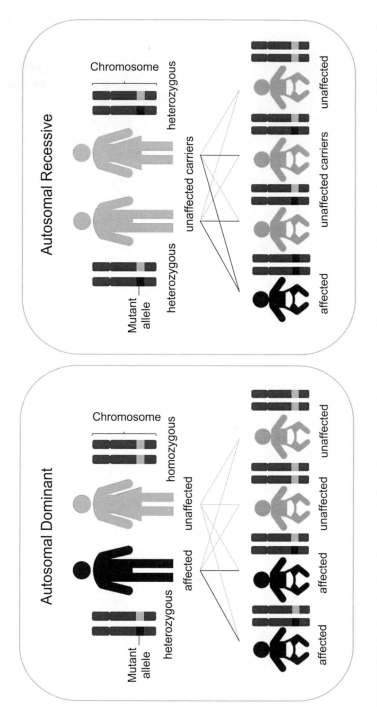

FIGURE 2-1 Genetic disorders encoded on the non-sex chromosomes (the autosomes) have dominant or recessive inheritance depending on whether one or both parental genes are required to harbor pathogenic variants for the disorder to occur. For a dominant condition, only one such variant is enough for the individual to be affected, while recessive conditions require pathogenic variants to be present on both copies of a chromosome pair.

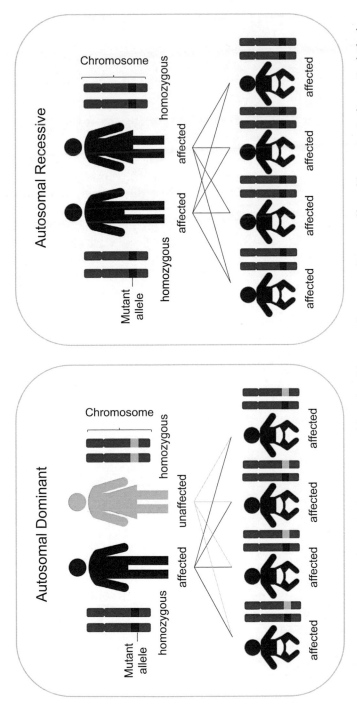

FIGURE 2-2 Circumstances in which parents would *not* be able to produce an embryo unaffected by a genetic disease include those in which one parent is homozygous for a dominant genetic disorder or both parents are homozygous or compound heterozygous for a recessive genetic disorder.

This report distinguishes between these two types of circumstances—those in which all of a couple's children would inherit the disease-causing genotype and those in which only some children could inherit it. As discussed in the next section, these latter couples currently have various options for having children lacking the disease-causing genotype.

Current Reproductive Options for Parents at Risk of Transmitting a Monogenic Disease

Over the past 30 years, a range of options has been developed to allow prospective parents to know whether they are at high risk of having a child who will suffer from a serious genetic disease and, if so, to avoid this outcome. An understanding of these options is important in assessing the circumstances in which HHGE might meaningfully improve or expand the options already available to prospective parents. Six current options are described below. Certain options may be acceptable to some prospective parents and not to others, while the availability of a particular option to a given set of prospective parents may also be constrained by cost, access, national regulatory policies, or other factors such as religious, cultural, or personal beliefs. Of course, a proportion of genetic diseases are the result of the *de novo* mutations discussed above, with the precise proportion varying by disease (Acuna-Hidalgo et al., 2016). Such mutations are unpredictable; therefore, the diseases that result from them are not amenable to prevention in that first generation of offspring using preimplantation or prenatal genetic testing, or HHGE were it ever to be available.

Preconception Genetic Testing

Some prospective parents know that they are at higher risk of having a child with a serious genetic disease because one of them has a genetic disease, because they have a family history of a genetic disease, because they underwent genetic testing for a targeted set of diseases that are at higher frequency in a particular ancestry group (e.g., Finnish or Ashkenazi Jewish individuals), or as a result of population genetic screening or testing.

Other prospective parents may not have access to family history information. And many parents only learn that they are at risk when they have an affected child; this is frequently the case for the thousands of rare recessive diseases. In populations with prevalent disease-causing founder mutations and/or high levels of consanguinity, preconception testing can enable prospective parents who wish to do so to reduce the risk of having children with serious monogenic diseases.

Adoption

Adoption avoids the risk of prospective parents passing on a genetic disease because the child is not genetically-related to either parent. Some people who would like to have children find that adopting a child is a positive and fulfilling way to create a family. Others would prefer to have a child who is genetically related to them.

Gamete and Embryo Donation

Another option is to conceive a child via egg or sperm donation, depending on whether the genetic disorder is likely to be transmitted by a woman or a man. They will experience the pregnancy and birth, and the child will be genetically-related to one parent (the father in the case of egg donation, the mother when sperm donation is used). Prospective parents may also use embryo donation. As with gamete donation, they will experience the pregnancy and birth, but like adoption, neither parent will be genetically-related to the child. The large proportion of fertility patients who seek treatment using their own gametes, such as intracytoplasmic sperm injection (ICSI), in preference to treatments involving donated gametes, such as donor insemination, illustrates the value placed on having genetically-related children. Nevertheless, many fertility patients who are unable to have genetically-related children come to accept the use of donated gametes or embryos.

Prenatal Genetic Testing

Some prospective parents have a strong desire to have a child who is genetically-related to both parents—that is, conceived from their egg and sperm. In the early phase of genetic testing, prenatal screening became available as an option to avoid having a child with a serious monogenic disease and is the method of choice for some people. The prospective parents choose to conceive a child in the conventional manner, have genetic testing performed on the fetal tissue (or the placenta in the context of non-invasive prenatal testing), and have the option to terminate the pregnancy if the fetus is found to be affected by the disease.

Preimplantation Genetic Testing

In the 1990s, another option became available: IVF coupled to PGT.[3] Developed in 1978, IVF made it possible to create a pregnancy by fertil-

---

[3] PGT for monogenic diseases is usually called PGT-M. There are other types of PGT, but for the sake of simplicity in this report we use 'PGT' to mean 'PGT-M', unless otherwise stated.

izing an egg outside of the body, allowing the resulting embryo to develop for a few days, and transferring it into a woman's uterus. PGT involves removing a few cells from an early embryo, identifying embryos that do not carry the disease genotype, and transferring one of those into the uterus (see Figure 2-3). IVF in conjunction with PGT is currently a reproductive option for many monogenic disorders. Boxes 2-1 and 2-2 discuss the processes involved, including potential harms and benefits, and current outcomes in terms of children born without a genetic disorder.

Treatment of Genetic Diseases

Finally, new options are emerging that would allow a child who is born with a serious genetic disease to be effectively treated. Our growing knowledge of the genetic basis of human disease is leading, in some cases, to therapeutics that can ameliorate or even prevent the serious effects of certain genetic diseases. Some prospective parents at risk of passing on a genetic disorder may choose to proceed to have children, depending on the effectiveness, accessibility, and affordability of the treatment options. The children with genetic disorders who are treated in such ways would remain at risk of passing on the disease to their own children.

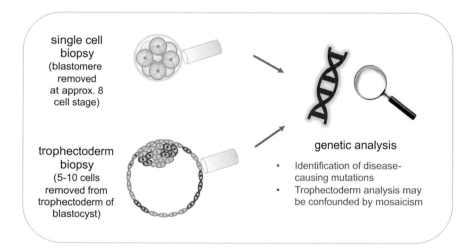

single cell
biopsy
(blastomere
removed
at approx. 8
cell stage)

trophectoderm
biopsy
(5-10 cells
removed from
trophectoderm of
blastocyst)

genetic analysis
- Identification of disease-causing mutations
- Trophectoderm analysis may be confounded by mosaicism

FIGURE 2-3 PGT involves the removal of genetic material at one of two different stages: (1) single-cell (blastomere) biopsy, or (2) trophectoderm biopsy in which several cells are removed from the blastocyst stage of the developing embryo. The genetic material (DNA) is then amplified and analyzed.

BOX 2-1
In Vitro Fertilization

Developed to help prospective parents who have difficulty conceiving a child, IVF involves the fertilization of an egg by sperm outside of the body. IVF is an intensive process that carries medical risks. Before starting IVF treatment, the patient or couple undergoes extensive screening. The woman then begins the IVF cycle, which takes 3 to 6 weeks. To induce ovulation, she is prescribed fertility hormones that stimulate the ovaries to produce multiple eggs. After 1 to 2 weeks of ovarian stimulation her eggs are typically ready for retrieval. The woman is carefully monitored during this period in order to extract as many eggs as possible while protecting against the development of ovarian hyperstimulation syndrome (OHSS). Mild OHSS causes abdominal swelling, discomfort, and nausea and occurs in up to 33 percent of women undergoing IVF. Just over 1 percent of all IVF patients develop moderate or severe OHSS, which can require admission to a hospital to treat the symptoms of vomiting and difficulty in breathing. The most serious potential complications of OHSS are blood clots, which can be fatal. The risk of OHSS is higher in women who have polycystic ovary syndrome, are under age 30, or who have had OHSS before (RCOG, 2016).

Following ovarian stimulation, 10 to 20 eggs are typically harvested and fertilized, either by mixing with sperm or by directly injecting a single sperm into each mature egg, a process known as ICSI. Following successful fertilization, development of the embryos is monitored for 2 to 5 days. Several high-quality embryos are typically then ready for implantation, and the embryologist chooses the highest-quality embryo to be transferred into the woman's uterus, most commonly on day 5 (the blastocyst stage). In some instances, multiple embryos are transferred simultaneously, although this practice is increasingly frowned upon by professional bodies as it increases the probability that the mother will have twins or triplets, which raises serious health risks for both the mother and babies. Patients can choose to freeze any extra embryos for later use, for example, in case the first cycle is unsuccessful or if the patient desires another child.

The success rate of IVF, measured in terms of live birth rate per embryo transferred, is typically in the range of 20 to 30 percent, depending on maternal age, embryo status, reproductive history, cause of infertility, lifestyle factors, and protocol used (the use of fresh or frozen embryos) (De Geyter et al., 2020; HFEA, 2018). The primary risks of IVF include multiple births, which carry a higher risk of early labor, premature delivery, and lower birthweight than single pregnancies; OHSS; miscarriage; complications in the egg-retrieval procedure; ectopic pregnancy; and stress.

IVF tends to place substantial physical, financial, and emotional burdens on the prospective parents. The cost of cycles of IVF may be covered by the health systems of some countries (as is the case in Israel, France, and the Netherlands), while in other countries or for certain types of couples (e.g., those with no known fertility impairment), they are not.

BOX 2-2
Preimplantation Genetic Testing

Prospective parents who know they are at risk of having a child affected by a particular monogenic disorder may decide to use IVF in conjunction with PGT, to ensure that embryos selected for transfer do not carry the disease genotype. The first successful use of IVF with PGT occurred in 1989, leading to the first birth in 1990 (Handyside et al., 1990).

Demand for PGT for monogenic diseases has been steadily increasing. PGT begins with the same processes as IVF, although ICSI is the much more commonly used fertilization method. Embryos are allowed to develop in an incubator for 3 to 5 days until they reach a stage where a small sample can be removed and tested for specific genetic diseases (see Figure 2-3). Performing the biopsy on day 3 at the cleavage stage allows the embryo to be transferred into the uterus earlier, while performing the biopsy on day 5 at the blastocyst stage allows for more genetic material to be analyzed. Depending on the day on which the biopsy is performed, between 1 and 15 cells are removed and subjected to genetic testing. As techniques of embryo culture and manipulation have improved, biopsies are increasingly done at the blastocyst stage, given the advantages of having more genetic material available for testing (Zanetti et al., 2019).

Because the genetic test looks for the presence or absence of a known pathogenic variant of a gene inherited from one or both parents, most PGT testing methods focus on analyzing a single locus or a localized region of the genome using sequence amplification by polymerase chain reaction, although sequencing methods that map the whole genome are increasingly being used (Zanetti et al., 2019). The test gives a clear answer on the presence or absence of the genotype in question in more than 90 percent of embryos biopsied.

Based on the test results, embryos are identified as either affected or unaffected. Unaffected embryos, if any, are selected for implantation. Following improvements in methods for freezing embryos, "selection of fresh embryos for transfer by PGT is increasingly being replaced by frozen embryo transfer." Freezing embryos "allows for more time to perform high-quality PGT and aggregate more 'diagnostic cases' for simultaneous examination, which also decreases costs" (Harper et al., 2018, p. 8).

If enough high-quality embryos are available for screening, PGT usually identifies unaffected embryos. When too few high-quality embryos are available, however, no unaffected embryos may be identified by the process or all of the identified unaffected embryos might be of low quality. In addition, some embryos may be damaged during the biopsy procedure, rendering them unusable or reducing the chances of a successful pregnancy.

For any given couple, the process of biopsy and genetic selection thus reduces the likelihood of success for IVF with PGT compared with IVF alone. From the European Society of Human Reproduction and Embryology (ESHRE) PGT consortium data, live birth rates per embryo transferred were 26 percent following PGT for single gene disorders. Broken down by the type of genetic disorder, the live birth rates per embryo transferred were 22 percent for X-linked disorders, 28 percent for autosomal recessive disorders, and 26 percent for autosomal dominant disorders (De Rycke et al., 2017). The ESHRE PGT consortium data for 2011 and 2012 also show that, of the PGT treatment cycles that reached the diagnosis stage (where at least one viable embryo had been produced), 80 percent resulted in an embryo transfer. In theory, HHGE might provide a new option in a proportion of those 20 percent of PGT cycles that reach the viable embryo stage but do not result in a transfer. It is not possible to tell how large that proportion is, because the data do not distinguish between cycles that were abandoned because they produced no unaffected embryos as opposed to those abandoned for any other reason, such as damage caused by the biopsy procedure.

More detailed data are available from individual clinics. Steffann et al. (2018) reported that, out of 457 treatment cycles of PGT in a 5-year period at the PGT Centre of Béclère-Necker hospitals in Paris, 72 cycles did not result in embryo transfer (n=50 couples), mainly because no unaffected viable embryo was available for transfer (52 cycles, n=43 couples) or because unaffected embryos stopped their development and failed to reach the blastocyst stage (20 cycles, n=18 couples). For this one clinic, 84 percent of the PGT cycles ended with a uterine transfer (slightly higher than 80 percent reported in the ESHRE data), and 11 percent of the cycles to PGT could theoretically have benefitted from HHGE, as all of the viable embryos were affected.

The ideal data to address the question of what proportion of couples do not manage to have a child following the PGT process would come from the analysis of success rates per couple rather than per treatment cycle. It is very hard to find such data on cumulative success rates per couple. For one PGT clinic in the United Kingdom in 2016, the live birth rate for couples with single-gene disorders was 39 percent per couple starting treatment, 54 percent per couple reaching transfer, and 70 percent when the couple had two or more unaffected embryos available (Braude, 2019). It would be valuable to have more systematic data about the overall success rates of PGT as a function of inheritance type, age, and number of cycles.

A number of ethical concerns have been raised about IVF in conjunction with PGT, which also could apply to HHGE. Some countries also allow PGT for social sex selection or for the selection of "savior siblings" who are genetically-compatible at the major histocompatibility locus with an existing child with a fatal disease and can provide an organ or cell transplant. These concerns are likely to be compounded by the advent of new sequencing technologies that enable the detection of "not only the genetic variants of interest, but also genomic variation unrelated to the original referral and request of the couple" (Harper et al., 2018, p. 8), which could lead to the selection of embryos based on genetic factors beyond the presence or absence of a specific disease-causing mutation.

Limitations of Current Reproductive Options

IVF in conjunction with PGT offers an option for many at-risk prospective parents wanting to have a child who is genetically-related to both parents and who does not suffer from a serious genetic disease. Two limitations, however, currently keep IVF with PGT from being a complete solution. These are circumstances in which all embryos produced by a couple would carry the disease genotype, and circumstances in which a viable genetically unaffected embryo is not identified through IVF and PGT cycles. HHGE has been suggested as a possible solution to these limitations. If clinically available, HHGE could also reduce the number of ovarian stimulation cycles a woman has to undergo before having a child, which would be of particular benefit to women at greater risk of ovarian hyperstimulation syndrome and those toward the end of their reproductive years.

Couples Unable to Produce Unaffected Embryos

In extremely rare cases, couples cannot produce any unaffected embryos. For these couples, the parental genotypes guarantee that 100 percent of their embryos will carry the disease genotype (see Figure 2-2). Such couples are extremely rare, because, in the case of an autosomal recessive disorder, both partners would be affected by having the disease-causing genotype in the same gene and would need to have reached reproductive age with a health status compatible with a pregnancy. In the case of an autosomal dominant disorder, one partner would be homozygous for the disease-causing mutation and would also need to have reached reproductive age and be able to produce viable gametes and if female, be able to sustain a pregnancy. With the advent of treatments for genetic diseases, it has been proposed that the number of such couples is likely to increase in the coming decades. For such couples, HHGE would represent a major new option because it could make it possible for the first time for them to have a child genetically-related to both parents but without the disease-causing genotype.

Couples for Whom Unaffected Embryos Are Unlikely to Be Obtained by Cycles of In Vitro Fertilization in Conjunction with Preimplantation Genetic Testing

For other couples at risk of having affected offspring, some fraction of their embryos will be genetically unaffected (e.g., an average of 50 percent in the case of one parent with an autosomal dominant disease and 75 percent when both parents are heterozygous for recessive disease mutations). For such couples, PGT provides a viable option for having a genetically unaffected child. If a sufficient number of eggs can be obtained from the female partner, it should be possible to identify and implant unaffected embryos. However, IVF followed by PGT sometimes fails to yield any unaffected high-quality embryos to transfer. Couples may choose to repeat the procedure, although some couples do not succeed even after several cycles. The current efficiency of IVF+PGT is described in Box 2-2. HHGE has been proposed as a strategy that might improve the current efficiency of IVF combined with PGT by genome editing of high-quality embryos that have the disease genotype, thus making them available for transfer (Steffann et al., 2018). Whether HHGE would provide a meaningful improvement in efficiency over existing protocols of IVF in combination with PGT is currently unclear and would depend on the extent to which it yields an increase in the number of embryos suitable for transfer.

Identifying the Genotype of a Zygote by Polar Body Genotyping

Current genome editing techniques would involve treating zygotes, at the single-cell stage, when it is not possible to determine their genotype directly without destroying the cell (see discussion in section "Heritable Genome Editing: The Use of Genome Editing in Zygotes," below). In the case of couples who exclusively produce zygotes carrying the disease-causing genotype, genome editing could proceed on all zygotes without risk of exposing genetically unaffected embryos to the potential harm of the editing machinery without potential benefit. In contrast, when couples can produce both genetically affected and unaffected embryos, subjecting all zygotes to editing would often subject unaffected zygotes to editing.

Polar body genotyping could, in certain cases, provide a reliable way of distinguishing zygotes that do and do not have a disease-causing genotype (see Figure 2-4). Polar bodies are cells produced as an oocyte progresses through the meiotic divisions. The developing oocyte reaches a stage in which it carries four copies of each chromosome, rather than the normal two. As it proceeds through meiosis, this number is reduced to one of each (a haploid set) that is combined with one copy of each chromosome coming from the sperm upon fertilization. The reduction is accomplished by expelling two sets of chromosomes into the first polar body (PB1) at meiosis I, prior to fertilization, and one set into the second polar body (PB2) at meiosis II, after sperm entry. Both polar bodies are accessible for analysis.

PB1 contains the two copies of each particular chromosome that were inherited from either the woman's mother or her father, selected at random. PB2 contains one copy of each of the chromosomes that were left in the zygote. Thus, analysis of the DNA in the two polar bodies reveals, by elimination, the alleles remaining in the zygote.

In the simplest case, when the woman is heterozygous for a disease-causing mutation, analysis of PB1 will show whether both copies of that mutation

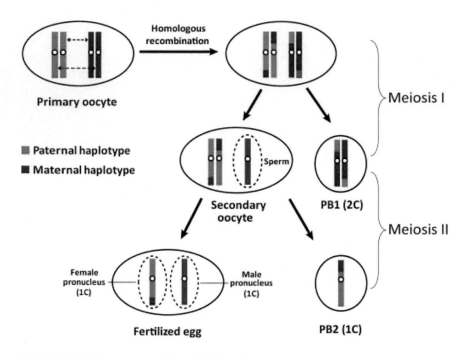

FIGURE 2-4 Formation of PB1 and PB2 during oocyte meiosis.
SOURCE: Reprinted with permission from Hou et al. (2013).

are present in PB1 or were retained in the oocyte. In the case of a dominant disease, if the oocyte has retained two disease-causing copies then any person that results from fertilization of that oocyte is certain to inherit that disease.

The situation is complicated by genetic recombination during meiosis I that can exchange a segment of each chromosome between parental copies prior to PB1 expulsion, in which case PB1 might show one copy of the disease-causing allele and one of the non-disease-causing allele. In this case, analysis of PB2 can resolve the issue of whether the zygote has received the disease-causing allele, since the remaining disease-causing sequence must be present either in PB2 or in the zygote. The exception to this would be if a gene conversion event has taken place that has changed the number of disease-causing alleles to one or three, but such events are rare.

In practice, determining whether PB1 carries two copies of the non-disease-causing allele is not entirely straightforward. Polymerase chain reaction (PCR)-based genotyping of PB1 is intended to detect the presence of an allele, but cannot reliably determine the number of copies present. It is possible that "allele dropout" (the failure of an allele to be detected) could cause a PB1 that is actually heterozygous to be mistakenly called homozygous for an allele. To avoid such errors, it will be important that PB1 be genotyped with a sufficient number of flanking genetic markers to ensure that the genotype at the disease-causing locus can be inferred with a high degree of certainty.

Polar body biopsy is a common and safe technique in PGT, used to detect maternally-derived chromosomal aneuploidies and translocations in oocytes (Schenk et al., 2018). The technique is also used for preimplantation diagnosis of monogenic diseases (Griesinger et al., 2009). However, because the paternal contribution to the genetic constitution of the developing embryo cannot be diagnosed by polar body analysis, its application remains limited (Altarescu et al., 2008).

For the purposes of HHGE, secondary oocytes that were diagnosed to have a disease-causing allele (because their associated first polar bodies had been shown to be homozygous for a non-disease-causing allele) could be frozen to provide a "reserve." These cells could potentially be used in an HHGE process, if all oocytes collected through the successive IVF attempts, and carrying at least one unaffected allele, had been used in a conventional PGT process but without success. This approach could ultimately increase the chances of a woman with autosomal or X-linked dominant disease having a healthy child without requiring a new IVF cycle.

For autosomal recessive diseases, the allele contributed by a mother heterozygous for a disease-causing allele could also be deduced by the same procedures. However, this would only allow inference that the zygote has biallelic mutations if the father had biallelic mutations.

GENOME EDITING:
SCIENTIFIC BACKGROUND FOR A TRANSLATIONAL PATHWAY

The success of therapeutic genome editing depends on both the clear identification of the disease-causing DNA sequence that needs to be changed and the reliability of the technical approach to accomplishing that change without undesired consequences. In this section, the current status of genome editing methods is reviewed, and existing limitations are highlighted. The focus is on the CRISPR-Cas platform, due to its prominence in research and in developing clinical applications, while the parallel utility of other platforms—zinc-finger nucleases (ZFNs) and transcription activator–like effector nucleases (TALENs)—is acknowledged.

Genome Editing Technologies

The modern tools of genome editing have contributed to a revolution in genetics because they provide the ability to introduce with relative ease specific, desired modifications at any locus, or loci, in the chromosomes of living cells. The idea of precisely editing the genomes of living mammalian cells dates back to the 1980s, when geneticists working with mice developed ways to use homologous recombination to introduce DNA into specific locations in the genomes of embryonic stem cells, which could then be used to create mice with desired genotypes (Capecchi, 2005; Doetschman et al., 1987). While workable for research purposes, the initial methods had very low efficiency, and the desired change was only made in a small number of the cells targeted. The secret to increasing the efficiency was the ability to introduce a targeted double-strand break at a unique, chosen target using a programmable nuclease (an enzyme that cleaves DNA). Various programmable nucleases, including mega nucleases, ZFNs, and TALENs, were successfully used (Bibikova et al., 2003; Joung and Sander, 2013). But the situation changed with a series of discoveries, over the course of two decades, culminating in the recognition that bacteria contain adaptive immune systems, called CRISPR-Cas, that are naturally programmed by ribonucleic acid (RNA) to cut specific DNA sequences and can be used to readily edit the genomes of living human cells (Doudna and Charpentier, 2014; Hsu et al., 2014; Karvelis et al., 2017).

Because of its simplicity and flexibility, the CRISPR-Cas platform has come to dominate research uses of genome editing, and it forms the basis of many preclinical studies and clinical trials (as well as applications in many animals and plants). The basic components of this platform are a Cas nuclease (Cas9 in the most widely used version) and a guide RNA (gRNA) that associate to form a complex. The gRNA usually consists of one RNA molecule (sometimes two) and provides specificity for the editing—directing the complex to a genomic DNA sequence (the target) that matches the variable portion of the gRNA. The gRNA associates with this DNA target through complementary base pairing. Typically, about 20 bases in the gRNA must match the target for effective recognition. Because of this length requirement, recognition can be quite specific, even in a complex genome like that of humans. Once a target is located, Cas9 cuts both strands of the DNA, leaving a double-strand break at that site. These breaks could be lethal to cells, but cellular mechanisms exist to repair them, providing the opportunity to change the DNA sequence at the target location (see Figure 2-5). The CRISPR-Cas system is highly flexible because the variable portion of the gRNA can be designed to match almost any desired target sequence. Although each Cas protein's enzymatic activity is restricted to a particular short sequence next to the gRNA-determined target, called the protospacer-adjacent motif (PAM), many Cas variants, both natural and derived, recognize different PAMs, and an appropriate one can be chosen for each specific target. In addition, the Cas-induced break can be made at variable distances from the site of the desired change and still be effective. Thus, it will be quite rare for any particular target to be inaccessible due

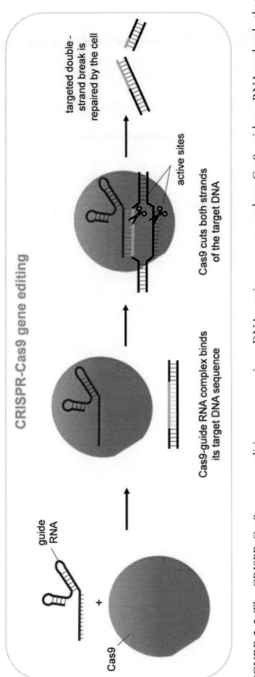

FIGURE 2-5 The CRISPR-Cas9 genome editing system pairs a DNA-cutting enzyme, such as Cas9, with a gRNA molecule that binds to the sequence of the gene to be edited. After the Cas9 protein cuts both DNA strands, the cell detects and repairs the double-strand break via any one of several different mechanisms.

to the absence of a suitable PAM. Even in these cases, the well-developed ZFN and TALEN technologies could complement CRISPR-Cas for editing these loci.

On-Target Modifications

Genome editing technologies rely on repair mechanisms in human cells to make the desired changes in DNA. As a result, the efficiency and specificity of genomic alterations depend not only on the properties of the genome editing system introduced into cells but also on the characteristics of the cellular repair mechanisms.

Cells have several mechanisms to repair the breaks that are created, each of which has advantages and disadvantages for making intended changes. One mechanism, known as non-homologous end joining (NHEJ), simply reconnects the broken ends. This process often results in the addition or deletion of DNA sequences (indels) at the site of the break (Rouet et al., 1994) (see Figure 2-6). Such changes can disrupt the normal function of the DNA at that site if it encodes a protein, for example, or governs the expression of nearby genes. If multiple breaks are made in a single cell, DNA can undergo rearrangements that can also have consequences for gene function. Although it is sometimes possible to anticipate new sequences that will be generated by NHEJ, it is not currently possible to control the

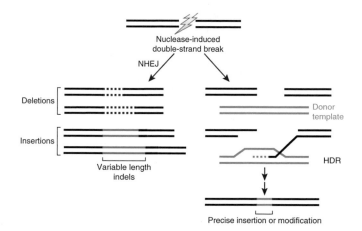

FIGURE 2-6 A cell uses two main mechanisms to repair a double-strand break at the targeted site. The most common is NHEJ, which often results in base insertions and deletions that can disrupt a gene. HDR uses a template DNA sequence to make more precise gene modifications.
SOURCE: Sander and Joung (2014), reprinted by permission from Springer Nature.

process or to specify a particular product. Thus, NHEJ is useful when the goal is to disrupt an existing DNA sequence but not when a specific editing outcome is needed.

The other major class of processes that repair DNA breaks in cells is homology-directed repair (HDR). In this case, a related (homologous) DNA sequence is used as a template from which sequences are copied at the site of the break (see Figure 2-6). The template can already exist inside the cell—on the sister chromatid or on the other parental allele—or it can be introduced into cells along with the editing nuclease. The sequence changes introduced from a template can be as subtle as changing one or a few base pairs or can involve DNA sequence insertions or deletions of hundreds or thousands of base pairs. With both NHEJ and HDR, changes are induced specifically at the site of the break made by the editing nuclease. The overall efficiency of total modification (NHEJ plus HDR) can be very high in some circumstances, but the outcomes are difficult to control. While HDR is a more versatile and precise repair mechanism and therefore more useful for genome editing, NHEJ is the dominant repair process in most human cell types, and HDR operates efficiently only during some portions of a cell's cycle of growth and division (Gu et al., 2020; Heyer et al., 2010; Hustedt and Durocher, 2017). The efficiency of HDR also varies widely among cell types for reasons that are not fully understood. Although non-dividing cells typically show very low levels of HDR, there is considerable variability among rapidly dividing cells of different types, and the responsible mechanistic differences have generally not been identified.

Approaches have been tried to enhance the use of the proffered DNA template by HDR. These include providing the template in various molecular formats, linking the template to the Cas9 nuclease or to the gRNA, and manipulating cellular DNA repair activities (Liu et al., 2019b). The improvements in most of these cases have been modest; HDR efficiencies do not approach 100 percent; and unintended products are still produced at some level. Encouragingly, some recent publications report improved efficiencies, including instances in mouse embryos (Gu et al., 2020). Continued research into cellular DNA repair processes will be needed to increase the efficiency and specificity of genome editing, particularly the efficiency of HDR.

Beyond small indels produced by NHEJ, larger sequence changes have been found at sites of induced double-strand breaks. These include extensive deletions (Kosicki et al., 2018), occasional insertions of DNA sequences introduced as intended HDR templates, and chromosome rearrangements. Such products are not readily detected by targeted PCR-based assays that are commonly used, so protocols must be designed explicitly to determine whether they are present.

Another strategy for precise modification that has been used experimentally is microhomology-mediated DNA insertion (Paix et al., 2017;

Sakuma et al., 2016). Because this leads to sequence additions rather than replacements, it will not be applicable for restoring common genomic sequences by introducing only one double-strand break in most cases. To adapt this approach to genetic replacements, two Cas-induced breaks would have to be made, which increases the likelihood of unwanted on- and off-target events.

It is worth noting that while NHEJ might be clinically useful for somatic genome editing, this cannot be said for HHGE, at least in its initial uses. In somatic genome editing, if an intervention derives a clinical benefit by introducing genetic alteration that breaks a gene or a regulatory element, this is acceptable even if the resulting DNA sequence is rarely if ever found in the human population because the change is limited to that tissue in that individual. However, this would not be acceptable for HHGE because the consequences of such genetic alteration in every tissue and at all stages of development could be expected to be deleterious in many cases. The change could also be inherited by future generations. For this reason, it is considered crucial for any initial uses of heritable genome editing to change a disease-causing allele to a common allele in the population that is known not to cause disease. This can only be done by HDR or other technologies that specifically change one DNA sequence into a specific desired sequence. This represents a critical issue for the future use of HHGE.

Off-Target Modifications

From the beginning of research on genome editing, concerns have been raised that at the same time that desired changes are made at the intended target sequence, changes could be introduced elsewhere in the genome. The ability to reduce the frequency of unwanted changes and the ability to detect off-target events when they occur have both progressed in recent years. For the CRISPR platform, specificity has been improved through testing various gRNAs for efficiency with a particular target and modifying both the gRNA and the Cas protein (Chen et al., 2017; Kleinstiver et al., 2016; Slaymaker et al., 2016). Similar advances have been made for the ZFN and TALEN platforms (Doyon et al., 2011; Guilinger et al., 2014). Frequencies of off-target mutagenesis below 0.01 percent at individual at-risk sites have been achieved in some cases.

Several methods exist for identifying non-target genomic sequences that may be at risk of cleavage by any particular genome editing reagent and for detecting and characterizing the off-target editing that occurs (Kim et al., 2019). Genome-wide screening using bioinformatics tools can help to identify genomic sites that are most similar to the target site and thus may be at risk of undesired editing. More useful are methods that identify sites of actual cleavage. Digenome-seq does this by cutting purified genomic

DNA and locating sites where cleavage has occurred by whole-genome sequencing (WGS) (Kim et al., 2015; Tsai and Joung, 2016). GUIDE-seq and DISCOVER-seq, by contrast, capture sites that are cleaved in living cells and subject them to DNA sequencing (Tsai and Joung, 2016; Wienert et al., 2019). Once these off-target sites are identified for a particular nuclease (particular Cas9-gRNA combination in the case of CRISPR), they can be tested by polymerase chain reaction amplification and targeted deep sequencing to see to what extent these off-target sites have been edited in any specific situation.

Unbiased WGS (see Box 2-3) can also be applied to detect off-target changes, but it has some limitations. Because all of the genomic DNA is being read, no individual site within the genome is read as many times as is the case with targeted deep sequencing. Therefore, low levels of mutagenesis can go undetected. There is an inherent error frequency in all methods of DNA sequencing, and there is a background of natural *de novo* mutations in cells, accumulated as cells grow and divide. Therefore, it is difficult to know which novel sequences are attributable to these effects, as opposed to genome editing. For assessing the genomes of a single cell or a few cells from early stage embryos, since only very small amounts of genomic DNA are available, the DNA must be amplified before it can be subjected to WGS using current technology. At present there is no unbiased method that can uniformly amplify all genomic sequences, although progress has been made in this direction (Chen et al., 2017; Hou et al., 2013).

Other Editing Approaches

Several genome editing systems have been developed that do not rely on the creation of double-strand breaks at the target site in DNA. The avoidance of double-strand breaks is acknowledged by many as ultimately desirable in genome editing given the unpredictability of the cellular response to these. In addition, since these approaches do not rely on HDR, they may be more effective throughout the cell cycle.

Base editing is an alternative approach that involves chemically modifying DNA bases at the desired target (see Figure 2-7). It relies on the specificity of Cas9-gRNA but uses a version that makes only a single-strand break or no break at all and is linked to a deaminase enzyme, resulting in the ultimate conversion of one base pair to another base pair at the targeted site. The tools of base editing are undergoing rapid development. Early experiments showed that sequence changes were often produced at off-target sites in DNA and even in RNA, in some cases at non-targeted sequences. Recent modifications of the base editing reagents have significantly reduced these unintended effects without significantly compromising on-target activity (Doman et al.,

BOX 2-3
DNA Sequencing

DNA sequencing refers to the determination of the order of base pairs in a segment of DNA. Three categories of sequencing procedures are particularly relevant to genome editing. Next-generation sequencing (NGS) procedures can determine billions of individual DNA sequences in parallel, providing a broad and/or deep view of the characteristics of a given sample. All of the sequencing techniques have some degree of error in their reads.

Sanger sequencing. This is a standard technique that determines stretches of hundreds of base pairs, typically from a product from the polymerase chain reaction technique or a molecular clone when only a single sequence, or a few variants of it, is expected to be present. It would be used, for example, to read the sequence around the intended editing target in each of the parental genomes to ensure that the editing reagents are properly designed.

Targeted deep sequencing. This technique would be used to evaluate the various editing products at both the intended target and at suspected off-target sites. Individual segments are amplified by polymerase chain reaction in fragments of several hundred base pairs. These fragments are subjected to NGS, generating thousands to millions of reads representing the different products generated at a single genomic location by the editing procedure.

Whole-genome sequencing. NGS can be applied to the DNA of an entire human genome of 6 billion base pairs. This would allow determination of whether the editing procedure had produced sequence changes anywhere in the genome, regardless of prior expectation. There are some limitations, however. Highly repeated sequences in the genome are difficult to analyze and are typically depleted from the sample before sequencing, and some types of DNA rearrangements are not reliably revealed. Recently introduced approaches that read long DNA strands continuously ameliorate these problems. At present, WGS in embryos is limited by the small quantities of DNA available because of limitations on the number of cells that can be extracted safely.

2020; Gaudelli et al., 2020; Grünewald et al., 2019; Richter et al., 2020; Yu et al., 2020). In addition to concerns about potential off-target events, current base editors can make only certain types of DNA sequence changes, specifically, transition mutations (changing C to T, G to A, A to G, or T to C) but not transversions (changing A to C or T, G to C or T, C to A or G, or T to A or G). According to Rees and Liu (2018), approximately 58 percent of human disease alleles are single nucleotide variants, 62 percent of which could be reversed with current base editors. As a result, roughly one third (35 percent) of known disease mutations could potentially be addressed using

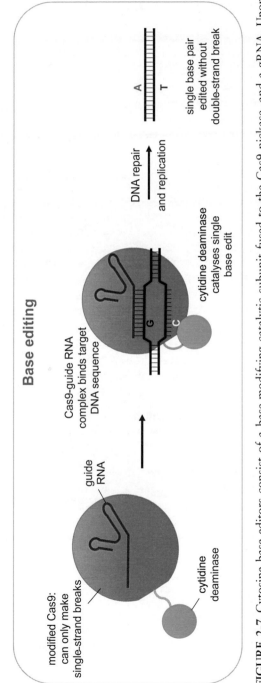

FIGURE 2-7 Cytosine base editors consist of a base-modifying catalytic subunit fused to the Cas9 nickase, and a gRNA. Upon binding the target DNA sequence, the helix is unwound and a single C base is converted to a U base. The unedited strand is cut by the modified Cas9, triggering the cell to repair the G-U mismatch to A-U, which becomes A-T upon DNA replication. Adenine base editors operate via a similar mechanism to convert targeted A-T pairs to G-C.
SOURCE: Rees and Liu (2018), adapted by permission from Springer Nature.

base editing technology. Nonetheless, the cytosine deaminase (C-to-T) base editor has been shown to be quite effective in human embryos, particularly at the two-cell stage (Zhang et al., 2019).

Very recently, novel base editors that induce C-to-A and C-to-G transversions have been reported (Kurt et al., 2020; Zhao et al., 2020). Currently these reagents also generate other products at the target, but they will undoubtedly be improved.[4]

Another recent innovation in genome editing is prime editing (Anzalone et al., 2019). This system involves modification of the gRNA that directs Cas9 to its target sequence such that the RNA also contains a repair template. The Cas9 protein is modified so that it cuts only one strand of the target DNA, in this case the strand that is not bound by gRNA. Cas9 is also linked to a reverse transcriptase enzyme that can utilize the extension on the modified gRNA to copy new sequences into the nicked strand. The provision of a template means that a much wider range of disease-causing mutations, including transitions, transversions, small insertions, and small deletions, can potentially be repaired when compared to base editing. Experience with prime editing is rapidly expanding (Sürün et al., 2020), and at least one study reports success in mouse zygotes, albeit at rather low efficiency (Liu et al., 2020).

While more research is required, both base editing and prime editing provide evidence of the flexibility of the CRISPR-Cas toolkit and the pace of ongoing development of precision genome editing methodologies. It is possible that continuing research may yield new methodologies that rapidly supersede the safety and efficacy of current editing approaches.

Non-Heritable Genome Editing: The Use of Genome Editing in Somatic Cells

One potential alternative to HHGE for the treatment of genetic diseases is somatic genome editing. This section discusses some of the relative advantages and disadvantages of somatic editing in comparison with HHGE.

The initial applications of genome editing in humans occurred in somatic cells, the cells that make up all of the cells of the body except sperm, eggs, and their precursor cells. The effects of genome editing carried

---

[4] It has also been proposed that genome editing could be used as an alternative to MRT to prevent the transmission of mitochondrial DNA (mtDNA) disease (Reddy et al., 2015). This study used mitochondrial targeted restriction endonucleases or TALENS and showed they could be used to potentially lower mutation load. However, this procedure led to net depletion of mtDNA and thus was not suitable for oocytes with a very high level of heteroplasmic or homoplasmic mtDNA mutations. A recent paper reports the use of base editing on mtDNA (Mok et al., 2020) and may represent a new approach to addressing mitochondrial disease. A detailed analysis of mtDNA editing requires a separate study, in the context of existing treatments for mitochondrial diseases and options such as MRT.

out in somatic cells are generally limited to the individual treated and would not be transmissible to that person's offspring. (The special circumstance of editing somatic cells that are located in an individual's reproductive system, such as editing in the testes to treat infertility, is discussed later in this chapter.) Despite the cost that would be associated with any clinical use of HHGE and the complex social, ethical, and scientific issues that heritable genome editing raises, the potential limitations associated with somatic editing, discussed below, represent one reason that HHGE has been proposed as a theoretical alternative for parents wishing to have a genetically-related child who does not have the disease-causing genotype.

Somatic genome editing is an option for treating patients with monogenic disorders, but it remains in early stages of clinical use, and much more experience will be needed to assess its safety and efficacy. The first clinical trial, initiated in 2009, tested the safety of using ZFNs to prevent the progression to AIDS in people infected by HIV (Tebas et al., 2014); and multiple trials using ZFNs, TALENs, and CRISPR systems are currently in progress.[5] With significant funding across multiple companies, somatic genome editing is likely to lead to numerous human trials in the coming decade.

The simplest targets for somatic editing are ones in which cells can be removed from a patient, treated outside the body, and returned (*ex vivo* genome editing) (Li et al., 2020). At present, the primary conditions that can be approached in this way are diseases resulting from mutations in HSCs. For example, promising results have been reported for patients affected with SCD and beta-thalassemia who were treated with CRISPR-Cas reagents to induce expression of fetal hemoglobin,[6] although long-term follow-up will be needed before conclusions can be drawn regarding its successes and limitations. Trials are also under way using genome editing to enhance the activity of CAR T cells for cancer immunotherapy (Bailey and Maus, 2019; Stadtmauer et al., 2020).

For many other envisioned somatic therapies, the genome editing reagents will need to be delivered directly to a patient's cells and tissues (*in vivo* genome editing). When a disease affects multiple organs, the challenge of delivery is magnified. Only in a few cases is the target tissue readily accessible. One favorable example is the eye, where direct injection of a viral vector carrying CRISPR-Cas reagents is feasible and is being applied for a rare retinal blindness condition.[7] The liver is also relatively accessible, and ZFNs are being employed to enhance a gene addition therapy in trials targeting hemophilia and metabolic disease.[8]

---

[5] See clinicaltrials.gov.
[6] See, for example, clinical trial numbers NCT03745287 and NCT03655678.
[7] See clinical trial NCT03872479.
[8] See clinical trial numbers NCT02695160, NCT03041324, and NCT02702115.

One feature of many of the above cases is that they rely on disruption of genome sequences by NHEJ. As noted above, this pathway is more active in most cells after a double-strand break is introduced than HDR. Treatments relying on HDR are in development, but attaining therapeutically relevant efficiencies remains challenging. For quite a number of genetic conditions, a non-disease-causing allele could be created via base editing and such approaches are being pursued actively.

While somatic genome editing avoids some of the challenging issues raised by HHGE—because somatic editing involves treating existing patients who can typically consent and because the resulting genetic changes would not be passed on to subsequent generations—somatic editing has some disadvantages. First, because editing does not alter the germline, a patient receiving somatic therapy for a genetic disease could still transmit the disease-causing mutation to future children. Additionally, because only a fraction of targeted cells might be edited, eliminating cells with the disease genotype or positive selection for the edited cells might be needed to increase the fraction of stems cells that have been edited. For example, protocols for somatic editing of hematopoietic stem cells (HSCs) commonly include cytotoxic chemotherapy to eliminate native HSCs before infusion of edited cells. These treatments confer risk of harm. Somatic genome editing therapies are also likely to be very expensive, although costs are unknown and likely to vary (Rockoff, 2019).

Heritable Genome Editing: The Use of Genome Editing in Zygotes

At present, the primary approach that could be used for undertaking HHGE would involve genome editing in zygotes. Because edits introduced would be present in every cell in the body, and the resulting genetic modifications could be passed on to subsequent generations, it would be critically important to obtain the desired genetic change at the target site and ensure an absence of editing-induced changes elsewhere in the genome. There are unique challenges in characterizing the editing events in zygotes and early embryos, as well as important gaps in understanding how to precisely control genome editing in these cells.[9]

---

[9] Genome editing technologies have also been adapted to affect the epigenetic state of somatic cells by altering DNA methylation (Kang et al., 2019) and histone modifications (Pulecio et al., 2017). Extensive epigenetic remodeling occurs during early development, and it is not clear whether epigenome editing would be heritable or how it would operate in zygotes and early embryos. Much more research on epigenome editing in embryos would need to be undertaken before it could be considered as an intervention for congenital imprinting disorders (Eggermann et al., 2015).

A zygote—the single, fertilized cell that results from the combination of parental gametes (the egg and sperm)—is the earliest stage of embryonic development. At first the maternal and paternal chromosomes remain in two distinct pronuclei in the cell, and then, after a round of DNA replication, they fuse to become a single nucleus. The zygote then divides into two cells, each with a nucleus containing the full complement of chromosomes from both parents. The blastocyst forms over the first week, and by day 7 consists of roughly 200 cells of 3 different types (Hardy et al., 1989; Rossant and Tam, 2017). Some cells, called the trophectoderm, are progenitor cells that will go on to form the placenta, while additional cells will go on to form the yolk sac. Approximately 10–20 cells within the inner cell mass (ICM) of the blastocyst are epiblast progenitor cells that will form the embryo proper (Niakan, 2019).

Most preclinical research on HHGE has focused on two techniques for getting the genome editing reagents (e.g., the Cas9 nuclease and gRNA, with or without a template DNA) into the zygote: (1) introducing them into an egg cell at the same time as the sperm, or (2) introducing them into the pronuclei or cytoplasm of the fertilized egg. Introducing these reagents can be done by direct mechanical injection or by electroporation, both of which have been used in human embryos without significant damage (Ma et al., 2017). At the zygote stage of development, only one copy each of the maternal and paternal chromosome sets are present, and the aim is to ensure accurate editing in these two chromosome sets while minimizing the chances of undesired consequences arising as a result of off-target events, mosaicism, or other issues. Some studies in mouse embryos have successfully used injection into two-cell embryos, where four or eight genomes are present (before or after S phase) (Gu et al., 2018; Zhang et al., 2019). This presents additional demands on the efficiency and consistency of the editing events in order to achieve uniform outcomes on all alleles and prevent mosaicism.

Another issue arises if editing is attempted in zygotes that have non-disease-causing genotypes, as would be the case if one or both parents were heterozygous for a recessive or dominant variant. To avoid making unnecessary edits in such embryos, the zygotes that require editing would have to be identified prior to treatment. The latter may be possible with polar body biopsy and genotyping (see above) or with an editing platform that can reliably edit a multicellular embryo following genotyping. This would depend on an editing protocol that can reliably edit an eight-cell embryo, the earliest point at which biopsy and genotyping is possible without harming the embryo.[10]

---

[10] Were HHGE ever to be used to the extent that a long history of safe use had provided confidence that there was nothing necessarily harmful about the process of introducing editing reagents into zygotes or early embryos, it might be considered acceptable to use editing

Efficiency of Editing at the Target Site

Efficiency of genome editing refers to the ability of the editing system to make the intended edit at the target site. To be used clinically, genome editing reagents would need to exhibit high efficiency in zygotes. First, the reagents must be very effective in binding to the intended target sequence. Second, the desired sequence modification must be produced with high efficiency.

Progress has been made in deriving Cas9 proteins and gRNA designs that yield essentially complete DNA cleavage of an intended target, including in human zygotes (Lea and Niakan, 2019). However, which DNA repair pathway will be used following DNA cleavage depends on cellular characteristics—including the stage of the cell cycle, which DNA damage response components are present, other factors that influence DNA repair, and potentially genetic background. Based on very limited experience, the process of HDR is not efficient in human zygotes. The more common result of making a double-strand break is the introduction of sequence insertions and deletions (indels) via NHEJ, and larger changes can occur as well (Kosicki et al., 2018; Lea and Niakan, 2019). This could result in replacing a disease-causing mutation with another mutation, the nature of which cannot be specified in advance. While the generation of indels has proved useful in the fundamental study of gene function during human embryogenesis (Fogarty et al., 2017), it would be a very undesirable outcome in clinical uses of HHGE. Several recent and not yet peer-reviewed preprints also report significant unintended editing near the target site in human embryos, including chromosomal modifications (Alanis-Lobato et al., 2020; Liang et al., 2020; Zuccaro et al., 2020). Further fundamental characterization of the process of DNA repair in early human zygotes and the development of effective strategies to facilitate use of the HDR pathway will be required to devise safe and effective solutions.

In mice, the introduction of the genome editing reagents at the G2 stage of two-cell embryos has been shown to improve rates of HDR (Gu et al., 2018); however, it is not yet clear whether the same applies to human zygotes. As noted above, editing at this stage would demand exceptional efficiency and consistency of the editing process to avoid mosaicism and to ensure that all alleles are edited. This was not achieved in reported mouse embryo experiments (Gu et al., 2018).

---

protocols that only targeted disease-causing alleles to treat a group of zygotes or embryos without prior identification of ones that carried the disease-causing genotype. This would depend on the development and experimental validation of genome editing reagents that are sufficiently specific that they modify only the disease-causing allele without affecting the non-disease-causing allele and did not introduce off-target modifications.

A recent report of high efficiency HDR in human embryos by gene conversion between maternal and paternal chromosomes at the target site is promising (Ma et al., 2017), but this interpretation has been challenged (Adikusuma et al., 2018; Egli et al., 2018; Ma et al., 2018), and further experimentation will be required for validation. Indeed, further research into DNA repair mechanisms and the possibility of gene conversion events in early human embryos will be critical.

Base editing and prime editing have been shown to generate very low levels of indels. For base editors, there is a window of several nearby base pairs in the target sequence that are at risk of unintended editing (Lee et al., 2020), although progress has been made in avoiding this outcome (Huang et al., 2019; Jiang et al., 2018; Kim et al., 2017; McCann et al., 2020). Technical developments continue to advance and will help address editing efficiency and specificity. A considerable amount of evidence has accumulated on the use of base editors in embryos, including those of humans (Li et al., 2017; Liu et al., 2020; Zeng et al., 2018; Zhang et al., 2019). Newer variants of Cas9 and gRNA systems and prime editing have not yet been extensively tested in embryos (Liu et al., 2020).

Beyond the questions of efficiency and repair pathways at the target site, there is also a possible issue with having to target two different disease-causing variants in prospective parents in whom one parent is compound heterozygous for mutations in the same gene that each cause a dominant disease or in which different alleles are present in the same gene in cases of recessive disease. If it is not possible to develop a single editing reagent that targets both variants, then there are two possible editing strategies, both of which have their limitations. These would be to (1) target one variant and use PGT to ensure that the resultant embryo had not inherited the other disease-causing variant, which increases the risk that there are no viable embryos without a disease-causing genotype available for transfer; or (2) introduce two editing reagents to target the two variants, which increases the risk off-target events and the possibility of chromosomal rearrangements.

There are further complications in using HHGE to prevent the inheritance of genetic disorders caused by the expansion of repeated DNA sequences, such as Huntington's disease. The challenge with such diseases is to reduce the number of repeats in the pathogenic copy of the gene to a non-pathogenic level, an approach that presents significant technical hurdles including the fact that there are identical triplet repeat sequences on both the other allele and elsewhere in the genome. One possible alternative editing strategy in such circumstances would be to introduce a stop codon into the disease-causing variant of the gene to prevent protein production, and therefore the pathogenic effect. Although such a strategy would prevent disease transmission, the precise DNA sequence that is produced would introduce a loss of function mutation into the population. In a population with high rates of consanguinity or small effective population size, in future generations this mutation could cause an increased frequency of disease due to inheritance of two copies of this mutation. For this reason, introducing heritable mutations that are potentially disease-causing in future generations are unsuitable for any initial uses of HHGE.

Specificity of Editing and Minimizing Off-Target Events

The specificity of genome editing systems—the ability to restrict activity of editing reagents to the intended site in the genome and not to make edits in undesired, off-target locations—is provided largely by the sequence complementarity between the gRNA and the DNA target in the case of CRISPR-Cas systems, or the protein-DNA recognition specificity in the case of ZFNs and TALENs. Before any use of genome editing in human embryos intended for establishment of a pregnancy, careful experimentation and optimization would need to be undertaken to select the reagents that would provide the greatest specificity in the genetic context of the prospective parents. Devising genome editing tools that induce only very low levels of off-target modifications in human zygotes appears feasible but would need to be validated in each specific case.

A second critical issue in evaluating unintended sequence changes in the zygote genome is the ability to detect with high confidence whether such changes have taken place. Tools have been developed for identifying sites that are at significant risk of cleavage by any particular editing reagent, for example, Cas9-gRNA combination (see discussion in section "Genome Editing Technologies," above). However, these methods have primarily been designed and tested in cell-free whole genomic DNA, in cultured cells, or in whole tissues or organisms, and they are not feasible in embryos, where there is limited availability of cellular DNA. Targeted sequencing of the off-target sites identified in cultured cells can be done with DNA from early stage embryos (Ma et al., 2017), but if there are sites that are uniquely at risk in zygotes, they will be missed.

Off-target sites shown to be at greater risk in somatic cell or embryonic stem cell editing with the same reagents can provide information to help guide the assessment of those sites in embryos. These experiments, however, will not be fully predictive of what occurs in a zygote, because the efficiency and specificity of editing are likely to vary with cell type (NASEM, 2017). As a result, the primary strategy for characterizing editing that has occurred in an edited embryo is WGS. To carry out whole-genome sequencing, a small number of cells are generally removed from an early embryo such as a blastocyst. Because the small amount of DNA is inadequate for processing, whole-genome amplification is undertaken prior to sequencing. This can introduce amplification bias, including allele dropout, where some sections of genomic DNA are amplified more efficiently than others and some not represented at all. Preclinical WGS of whole-embryo DNA could be useful in identifying the zygote-specific off-target sites, and it would not be as subject to the problem of small amounts of DNA. It would also be useful to assess whether there is a particular somatic cell analysis of off-target sites that correlates well with identification of such sites in zygote editing. Another issue with sequence analysis is that it must cover the full range of likely alterations that could have occurred. This includes large insertions and deletions and even whole or partial chromosome losses that are difficult to detect with standard procedures (Kosicki et al., 2018). Current methods thus lack sufficient power to locate and characterize off-target editing in early embryos with sufficiently high confidence.

Assessment of Mosaicism

Genome editing of embryos is performed as early as possible—generally at the single-cell stage—to maximize the chance that maternal and paternal genomes have been edited before significant DNA replication and cell division take place. If editing continues beyond this stage, different cells in the embryo may carry different sequence changes at the intended target or at off-target sites. This results in mosaicism, a condition that has been commonly observed in mouse genome editing experiments (Mianné et al., 2017). Mosaicism is a serious concern because some cells in the growing embryo would have the intended sequence change while other cells would not (see Figure 2-8). Unedited cells could make a significant contribution to tissues or cell types that contribute to disease causation, thereby undermining the disease prevention strategy. In addition, editing activity that continues past the single-cell stage raises the prospect of continuing off-target mutagenesis. The effects that genetic mosaicism of this type might have on development and post-natal life are difficult to predict but could be significant.

Preventing mosaicism requires a very high efficiency of the desired on-target modification in the one-cell zygote and restriction of editing activity beyond that stage. Research in somatic cells suggests that the Cas9-gRNA complex is rather short-lived, but the lifetime of the complex in human embryos has not been well characterized. New methods to restrict the duration of active editing are needed. For example, it may be possible to reduce the time the complex remains active by fusing Cas9 to protein domains that accelerate its degradation.

Mosaicism poses particular challenges for verification that correct genome editing has occurred in a clinical context. For an embryo destined

70

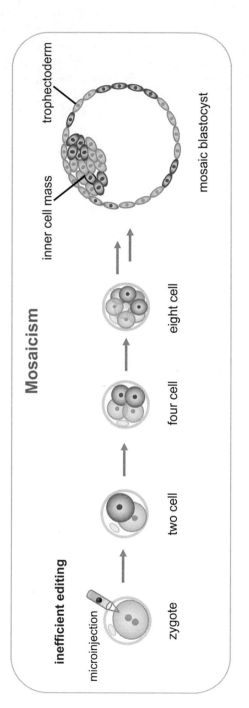

FIGURE 2-8 Mosaic embryos in which cells contain different genetic material can arise, for example, when edits occur in only one cell at the two-cell stage.

for transfer, only one blastomere or a few trophectoderm cells can be removed for molecular analysis. On- and off-target sequence analysis of these cells does not provide information on the genotypes of the remaining cells of the embryo, including the cells in the inner cell mass that will form the embryo proper. Another method under development is analysis of DNA found in fluid within the hollow central cavity of a blastocyst or in embryo culture media (Leaver and Wells, 2020). Further research is needed to address questions such as the source of detectable cell-free DNA—whether it results from random cell loss or is lost preferentially from cells that may have other developmental anomalies (e.g., aneuploidy). The continued development of non-invasive, cell-free DNA-based techniques may provide further options for the genetic characterization of human embryos prior to clinical use. However, such cell-free methods still could not provide information on the genotype of each cell in an embryo and therefore cannot guarantee the absence of mosaicism.

No current non-destructive method can determine whether all cells in the embryo carry exactly the same edits; it is difficult even to envision a method that could do so. For this reason, it will be essential that preclinical research on human zygotes establish procedures that only very rarely lead to mosaic embryos.

Assessing Early Embryonic Development: The Epigenome and Transcriptome

A number of additional assessments are important in characterizing the impact of genome editing in zygotes. One critical issue would be whether edited zygotes proceed through subsequent steps of development in a normal fashion. Following fertilization, embryos undergo a sequence of carefully orchestrated events that include and depend on epigenetic modifications.[11] Studies of preimplantation mouse embryos show that there are global changes to methylation of the maternally and paternally derived genomes, while histone proteins undergo modifications that alter nucleosome positioning and DNA accessibility to the cell's transcriptional machinery (Eckersley-Maslin et al., 2018; Li et al., 2019; Xu and Xie, 2018). Genome organization at higher levels also occurs in early embryos, including the formation of 3-D topologies that can enable DNA sequences separated by large distances to interact (Flyamer et al., 2017). The epigenetic remodeling that takes place in human embryos is likely to be more complex than in mice, due to the inbred genome of laboratory mice and the influence of genetic variation in humans on epigenetic variation (Delahaye et al., 2018).

---

[11] An epigenetic modification is one that can result in a change in gene expression without changing the DNA sequence of a gene.

Studies of the epigenomics of early human embryonic development are ongoing (Gao et al., 2018; Guo et al., 2014; Liu et al., 2019a; Smith et al., 2014; Wang et al., 2019; Zhou et al., 2019).

It is currently unclear whether making and repairing chromosomal breaks in zygotes, as might occur with HHGE, could have an impact on local and global DNA methylation, histone modifications, or chromatin domain organization. Preclinical research on epigenomic and transcriptomic profiles is needed to determine whether epigenomic characteristics and patterns of gene expression are altered in edited embryos compared to those that are untreated. Assessment of local chromatin dynamics in the vicinity of an edit may also be important. Research in genome-edited model organisms (e.g., mice) is likely to shed light on various molecular events requiring assessment in human embryos.

Methods for assessing some of these features at the single-cell level are available. For example, DNA methylation profiling of individual cells in the human preimplantation embryo has been reported (Zhu et al., 2018), and single-cell multi-omic approaches surveying chromatin state, nucleosome positioning, and DNA methylation are being developed (Li et al., 2018). Data emerging from stem cell–derived embryo models (Moris et al., 2020; Simunovic and Brivanlou, 2017) might also inform characterizations of early human development. Further fundamental research and refined assessment methodologies will be needed to establish whether developmental milestones, including epigenetic and transcriptomic profiles, are comparable to those of unedited human embryos.

Sources of Relevant Information to Inform Potential Development of a Translational Pathway

The evidence base that would need to be assembled for any clinical translational pathway for HHGE would need to draw on information obtained from a variety of sources.

In Vitro Systems

Most of the current information on editing of mammalian genomes comes from research in cultured cells, and these systems continue to provide valuable insights. Guidance for germline editing, at least for the foreseeable future, will come from methods for optimizing combinations of Cas9 and gRNAs, from the development of novel reagents such as base editors, and from testing various configurations of template DNAs for restoring non-disease-causing sequences. Additional methods for identifying and minimizing off-target mutations will need to be developed, including approaches for assessing the production of large insertions, deletions, and rearrangements.

Of particular importance are methods for controlling the editing outcome at the intended target, including suppressing indel formation and enhancing sequence replacement. Relevant results can be obtained from many different types of cultured cells, but perhaps most useful will be experiments in cell lines, induced pluripotent stem cells (iPSCs) and primary cells, carrying the particular disease-causing mutation. A limitation to the use of stem cells for this purpose is the fact that there is considerable variability among lines established from individual patients, and there are reports of high rates of genomic abnormalities (Henry et al., 2019).

Clinical Use of Somatic Genome Editing

Despite the fundamental differences between genome editing in somatic cells and in germline cells discussed above, clinical experience with somatic genome editing can provide some information to help inform HHGE. The differences in methodologies mean that successes or failures in the use of somatic genome editing are unlikely to have *direct relevance* to the prospects for HHGE in humans—the editing reagents and delivery methods used will likely differ, as will the cellular context and repair mechanisms in zygotes versus somatic cells.

Nevertheless, clinical trials of somatic editing therapies represent the use of genome editing in human primary cells, rather than in cells maintained in laboratory culture. Somatic therapy may thus be able to provide some insight into the benefits to a patient from correcting a particular disease-causing mutation, and it may also provide some data on the types and frequencies of genetic modifications observed at the intended target and at off-target locations in different cell types. Long-term monitoring of patients who receive somatic genome therapies may also reveal the extent to which there are late-appearing risks associated with the treatment, particularly if off-target mutations were created, or may help reveal whether therapeutic outcomes are influenced by the broader genomic background of patients. At the same time, populations of treated somatic cells undergo competition for growth and survival in the body, which might mask deleterious effects that could affect the development of treated embryos. Long-term follow-up of somatic therapy patients will also provide insights into compliance with long-term follow-up processes and optimization of communication and consent procedures.

Research in Zygotes of Other Mammals

While it is likely that human zygotes harbor capabilities and undergo processes that differ in significant ways from those in human somatic cells, it seems likely that human zygotes may share some of these capabilities and processes with zygotes of other organisms, particularly other mammals. Germline genome editing is being employed in many animal species for the purpose of generating specific variants for research purposes or for improvement of livestock traits. In these cases, there is no need to achieve a very high efficiency of the desired edit, since multiple individual animals can be screened at many stages of development and the final examples are often obtained in subsequent generations through selective breeding. Such approaches would be unacceptable in humans.

A great deal of genome editing research has already been done in mouse embryos, but, while undeniably useful in identifying key parameters, these may not be the best model for human embryos. While human and mouse blastocysts show similar morphology, there are significant differences at later stages of embryonic development. In addition, it has been shown that the timing of expression and the early embryonic function of certain genes differ significantly between the two species (Fogarty et al., 2017; Niakan and Eggan, 2013). Recently, more attention has turned to larger mammals, including cows and pigs, in which embryo genome editing is becoming routine. Work has also been done in non-human primates, including macaques and rhesus monkeys, which are the closest equivalents to humans (Chen et al., 2015; Niu et al., 2014). In some cases, human disease alleles may have been introduced into the genomes of these organisms, creating so-called humanized genes, and these would constitute excellent models for assessing the feasibility of specific sequence repairs. However, such work in larger mammals may be more likely to give rise to ethical objections in some countries due to the relative sophistication of the animals in question.

Attempts to perfect human germline genome editing can benefit greatly from focusing research in other zygotes on the issues that are currently most troublesome. These include raising the efficiency of on-target editing and preventing unintended on-target events. This will likely require developing an understanding of how DNA repair processes operate in zygotes, how outcomes depend on the timing of introducing the editing reagents, and what formats of template DNA and delivery are most effective. Off-target events can also be addressed. The issue of embryo mosaicism is critical and cannot be addressed in somatic cells. Methods to limit editing to the one-cell stage as well as methods to analyze mosaics can be developed in model organisms before being tested in human zygotes. The issue of effects of the editing procedure per se on epigenetic programming and gene expression can also be investigated. Of course, other issues, such as off-target effects, would not reflect the vulnerabilities in the human genome.

Research in Human Zygotes

Early human embryo development is already studied for a variety of reasons, including to provide insight into infertility, implantation, and placental development, and to improve IVF technologies. Currently, little is known about how the very first cell types that emerge in a human embryo become specialized in their fate and function (Lea and Niakan, 2019) or how this might be affected by genome editing.

Before it would be appropriate to proceed with HHGE, the necessary features of reliable germline genome editing would need to be tested in human zygotes directly. These tests must demonstrate very efficient introduction of the intended sequence change, no significant level of off-target mutagenesis, and a very low probability of mosaicism, and these tests must, at least for the earliest cases, be performed for each new target and Cas nuclease/gRNA combination. Obtaining meaningful measurements of the on-target, off-target, and mosaicism rates would require analyzing a sufficiently large number of embryos.

This will present a challenge, however, because research conducted on human embryos is prohibited in some countries and is subject to stringent oversight and regulation in many others. Where research is permitted, generally only a limited number of human embryos can be obtained, and these are generally donated surplus embryos from IVF. Many human embryo genome editing studies have also used non-viable tripronuclear embryos (Li et al., 2017; Liang et al., 2015; Tang et al., 2018; Zhou et al., 2017), but their abnormal chromosome content and aberrant developmental course make them unsuitable for preclinical characterization of genome editing in human zygotes as part of a potential translational pathway to HHGE.

With respect to adverse developmental effects, it is currently prohibited to experimentally monitor human embryonic development beyond 14 days because of the culture limit that exists in many jurisdictions (Cavaliere, 2017). Synthetic embryo-like entities may be of some use here because they can be maintained beyond 14 days; however, current models do not fully reflect the cell types, environment, or developmental timings of an intact embryo (Aach et al., 2017; Moris et al., 2020; Rivron et al., 2018; Warmflash, 2017). Monitoring fetal development requires implantation to establish a pregnancy. Therefore, experiments in non-human organisms must provide assurance that genome–editing-induced adverse developmental effects are likely to be non-existent or minimal.

FUTURE ISSUES IN ASSISTED REPRODUCTION: IMPLICATIONS OF IN VITRO STEM CELL– MEDIATED GAMETOGENESIS

Rather than undertaking genome editing in zygotes, an alternative pathway for HHGE would be through genome editing of cells that are capable of forming functional male and female gametes (sperm and eggs). Genome editing is unlikely to be undertaken directly in sperm or eggs, but several types of cells can serve as gamete precursors in which genome editing could be performed, including (i) spermatogonial stem cells (SSCs) and (ii) patient-derived pluripotent stem cells or nuclear transfer embryonic stem cells (ntESCs) that could be induced to differentiate into functional haploid gametes via in vitro gametogenesis (IVG). The technologies to develop human gametes from cultured stem cells are still under development and are currently unavailable for clinical use. Prior to any human use of such stem cell–mediated methodologies, they would need to be permitted in their own right as a component of assisted reproductive technology (ART). However, the development of in vitro stem cell–mediated gametogenesis would have significant implications for both the reproductive options that might be available to prospective parents and for HHGE.

Implications of In Vitro Stem Cell–Mediated Gametogenesis for Heritable Genome Editing

In the vast majority of cases, in vitro stem cell–mediated gametogenesis would eliminate a need for heritable genome editing as a means of preventing the transmission of monogenic diseases. For those circumstances in which a couple could produce an embryo without the disease-causing genotype, the ability to screen a large number of embryos created from male and female gametes produced from the prospective parents' somatic cells would enable suitable embryos to be identified. This would be the case even in those circumstances in which parents have a relatively low predicted chance of producing an unaffected embryo using conventional ARTs. This technology would thus eliminate the current efficiency issues associated with PGT.

The exception that would remain would be circumstances in which *all* embryos that could be produced by a couple would carry the disease genotype (see Figure 2-2). For such cases, HHGE would still be the only option for producing an unaffected child genetically-related to both parents. In this circumstance, IVG offers the potential to address a number of the technical challenges associated with using current methodologies to undertake genome editing in human zygotes. Performing genome editing in cultured cells would permit very careful analysis of any edited genomes at both genetic and epigenetic levels *prior to* the production and use of gametes. This would have significant safety implications, since the issues of on-target editing fidelity and avoidance of off-target events could be largely settled before any gamete is considered for use in the creation of an embryo. Meanwhile, high-throughput epigenetic analyses can be employed to ensure the epigenetic stability of genome-edited cells. In addition, the use of edited gametes to generate an embryo would avoid the issue of mosaicism, since all embryonic cells would be generated from a single gamete that had previously been genome-edited.

Approaches involving the use of human gamete and precursor cells in culture for laboratory research show promise for addressing basic questions in early human development. Any future clinical use of IVG, however, raises numerous scientific and ethical issues that would require careful consideration given the potential consequences for human reproduction (Bredenoord and Hyun, 2017; Greely, 2018). What would societal attitudes be to the routine production of hundreds or even thousands of human embryos for use in research or in treatment? Research would no doubt flourish in such circumstances, but in the context of treatment, would thousands of embryos be routinely discarded, or stored indefinitely, because some patients feel that destructive embryo research is ethically unacceptable? Some practitioners might attempt to screen thousands of embryos for polygenic traits (see the next section), ranking them according to polygenic risk prior to transfer (Karavani et al., 2019). As a result, wide-ranging societal discussions would need to occur prior to any clinical use of IVG, analogous to the discussions that are required prior to clinical use of HHGE (Adashi et al., 2019).

Preclinical Research Using In Vitro Stem Cell–Derived Gametes

The current state of progress in generating gametes from stem cells cultured *in vitro* is important when considering the potential implications of this technology for HHGE. It is unclear at present which, if any, of these approaches might reach a stage of development at which clinical application could be considered.

Genome Editing in Spermatogonial Stem Cells

Sperm cell genomes cannot be edited directly using current technology. However, sperm cells originate from stem cells in the seminiferous epithelia of the testis, the SSCs. SSCs can be isolated from multiple species, including primates, but so far have only been maintained long term in culture from small mammals (Kubota and Brinster, 2018). Research in mice has shown that genome editing of SSCs and their subsequent transplantation into the testis results in the production of sperm with the edited genome. This method could be used to prevent human genetic disease inherited from the male lineage. Wu et al. (2015) used the CRISPR-Cas9 system to edit the genome of mouse SSCs to correct a cataract-causing mutation. They were able to identify and select SSCs carrying the desired genome editing but lacking other unwanted genomic changes or signs of epigenetic abnormalities (including abnormal genomic imprinting), and were able to produce healthy offspring after transplantation of the edited SSCs back to the mouse testis. As with zygote genome editing, for any initial human uses to produce gametes, it would be very important to carry out investigations of the epigenetic and transcriptomic properties of embryos generated from edited SSCs. It is currently unclear whether the requirement to transplant cells into the testis would be an obstacle to clinical application or whether maturation into functional gametes could be reliably and safely accomplished *in vitro*.

Use of Androgenetic Haploid Embryonic Stem Cells

Research in mice also shows that androgenetic haploid embryonic stem cells (AG-haESCs) can be derived from embryos generated either by injecting sperm into oocytes from which the maternal chromosomes have been removed or by fertilizing eggs and removing the female pronucleus. These AG-haESCs, with genetic modifications to mimic the imprinted state of two paternally-imprinted genes, can be injected into oocytes to "fertilize" them, giving rise to live and fertile mice (Wang and Li, 2019). Therefore, genetic manipulation of AG-haESCs is a way of performing one-step transmission of genetic modifications in mice.

Human AG-haESCs have recently been successfully derived. These cells exhibit typical paternal imprints and can also "fertilize" human oocytes and support early embryonic development, leading to blastocysts and diploid embryonic stems cells with transcriptomes comparable to those of normal diploid embryos and embryonic stem cells, respectively, derived from ICSI (Zhang et al., 2020). Haploid embryonic stem cells thus provide a novel form of human germline stem cell that could potentially be used for editing disease-related mutations and validating the desired genotype.

Use of Induced Pluripotent Stem Cells or Nuclear Transfer Embryonic Stem Cells for In Vitro–Derived Gametogenesis

Genome editing could also be performed in patient-derived iPSCs or ntESCs, prior to these being differentiated into gametes *in vitro*— known as IVG (see Figure 2-9). Mouse pluripotent stem cells can be converted into cells with properties similar to primordial germ cells (called primordial germ cell–like cells, or PGCLCs) (Hayashi et al., 2011). When these are introduced into germ cell–free mouse gonads, functional spermatozoa can be produced. Further differentiation *in vitro* into germline stem cell–like cells and functional spermatid-like cells has also been reported (Ishikura et al., 2016; Zhou et al., 2016). These studies involve the completion of gametogenesis *in vivo* by gonadal transfer or *in vitro* by co-culture with neonatal testicular somatic cells but establish the principle that mouse stem cells could be converted into functional male gametes.

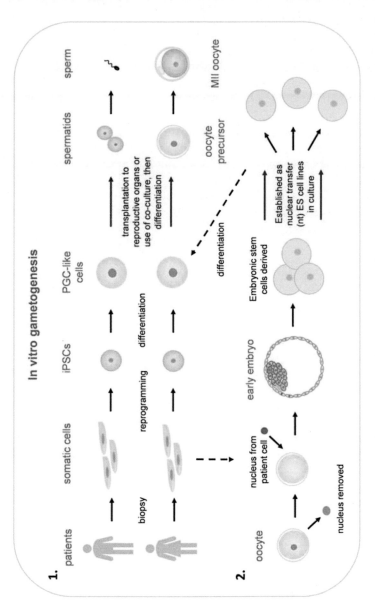

FIGURE 2-9 (1) Treating somatic cells from a patient with transcription factors/growth factors can reprogram the cells to become iPSCs.. (2) Alternatively, human embryonic stem cells can be derived from embryos following nuclear transfer into enucleated oocytes (ntESCs). These pluripotent cells can be differentiated into PGCLCs. The development of human PGCLCs into gametes could then be achieved by transplantation into reproductive organs or through the development of novel co-culture methodologies.

The ovary does not contain female germline stem cells analogous to SSCs that could be extracted and manipulated in cell culture, because gametogenesis is completed before birth in females. The only theoretical route to developing female gamete precursors for cultivation and genome editing is therefore through the use of IVG. Hayashi et al. (2012) reported that transplantation of PGCLCs into the ovaries of adult mice ultimately resulted in the production of fertile offspring. Morohaku et al. (2016) reported the successful *in vitro* maturation of mouse primordial germ cells into MII oocytes (the stage at which an oocyte would be fertilized by a sperm). Hikabe et al. (2016) reported the reconstitution of the entire process in mice, involving conversion of stem cells to PGCLCs and maturation of these PGCLCs into MII oocytes. These studies demonstrate that it is possible in mice to recapitulate the female gametogenesis pathway from pluripotent stem cells.

Studies of IVG using human cells are ongoing, but it is unclear whether it will be possible to replicate the successes reported in mice, especially the initiation and completion of meiosis. Human PGCLCs (hPGCLCs) have been derived from iPSCs (Sasaki et al., 2015) and characterized molecularly (Chen et al., 2019). One challenge to human female gamete derivation *in vitro* is the successful use of hPGCLCs. Yamashiro et al. (2018, 2020) reported the use of hPGCLCs to derive oogonia-like cells in a long-term culture model. However, this is not an efficient approach for the further differentiation of such in vitro–derived oogonia into primary oocytes in meiotic prophase I. Improved co-culture methods will likely be required, but the availability of fetal gonadal somatic cells of the appropriate type may be challenging. The use of such approaches to produce human sperm has not yet been achieved although *in vitro* reconstitution of some steps in spermatogenesis have been reported (Nagamatsu and Hiyashi, 2017; Yuan et al., 2020).

All IVG approaches would involve substantial periods of cell culture, and the adaptation to culture itself runs the risk of introducing undesired genetic or epigenetic changes. Further research would be required in mammalian models, including non-human primates, to develop this as a potential method of producing human gametes.

ADDITIONAL COMPONENTS OF ANY CLINICAL TRANSLATIONAL PATHWAY FOR HERITABLE HUMAN GENOME EDITING

In addition to the scientific and technical considerations discussed above, any potential clinical use of HHGE would entail the incorporation of detailed plans for obtaining informed consent and for monitoring the effects of genome editing. Lessons for developing such plans can be drawn from experiences with other novel human ARTs, such as mitochondrial replacement techniques (MRT), and from current practices in reproductive medicine, although HHGE would pose additional unique challenges due to the risks associated with making heritable changes.

Informed Consent

The potential use of HHGE for monogenic diseases poses specific challenges in terms of informed and voluntary consent that are analogous to those presented by MRT. These challenges are discussed in detail in reports from the Nuffield Council on Bioethics (NCB, 2012) and the National Academies of Sciences, Engineering, and Medicine (NASEM, 2016). HHGE for other types of uses, such as for polygenic disorders (see below), would raise additional challenges.

Prospective parents, as participants in initial human uses of heritable genome editing, would need to understand the novel procedures to be employed in addition to the use of IVF and PGT, and would need to be aware of the lack of information beyond preclinical evidence on the safety and efficacy of HHGE in humans in order to weigh the potential harms, benefits, and uncertainties involved in their decision on whether or not to proceed. Prospective parents would also need to be aware of the risk that they may give birth to a seriously ill or disabled child, and of the possibility that they may be faced with the difficult decision of whether or not to terminate a pregnancy should prenatal testing identify genetic or physical anomalies. The advantages and disadvantages of alternative routes to parenthood that would avoid the transmission of genetic disease would need to be carefully discussed as part of the consent process. It will also be important to discuss with the prospective parents the pressures they may face from media attention or public interest both during pregnancy and after the birth of a child with an edited genome.

Furthermore, prospective parents would need to be informed of the importance of monitoring of the health status of such children to document outcomes of HHGE and informed that they would be asked to consent to prenatal and long-term assessment of their children who have undergone an editing procedure. Informed and voluntary consent from parents for their children to participate in monitoring would need to be obtained at each new phase of assessment until the children reach the age of consent (generally 18, though this varies by jurisdiction). This consent would need to inform parents of all the features of the assessment that may affect their willingness to allow their child to participate and also of their right to refuse or withdraw from participation without incurring any penalty to themselves or their child. Parents would also need to be informed about the purpose of the monitoring; who would be conducting evaluations; what their participation would entail, including the risks, benefits, and limitations; and the safeguards that had been put in place regarding confidentiality, anonymity, and data protection.

The children themselves would not be able to give consent to long-term monitoring at the outset, as is the case for the many longitudinal studies that exist worldwide on children's health and development. Before they reach the age of consent, and once they are old enough to do so, children would be asked to assent, which means that they agree to participate without necessarily understanding the full significance of the assessment.

In line with guidelines from the Society for Research in Child Development (SRCD, 2007) and the report *Ethical Research Involving Children* (Graham et al., 2013), assent or informed consent would need to be obtained from children born following HHGE according to the children's age and developmental level. Children would need to be informed of the nature of the assessment in a way that is appropriate for their level of understanding, and assessors would need to ensure that children understand what participation will entail. It would also need to be made clear to children that they are free to participate or not and are free to withdraw, entirely or from specific assessments, at any time without having to give a reason and without adverse consequences. Children would also need to be assured of medical confidentiality and personal privacy. The people involved in the assessment of children must be trained to recognize signs of discomfort and instructed to cease the assessment should a child become distressed.

Due to the importance of monitoring children born following HHGE, every effort would need to be made to encourage parents and children to participate in long-term follow-up. For individuals with edited genomes who continue to consent to and engage in a monitoring process into their child-bearing years, this process would provide an opportunity to invite these individuals to include any children they have in an intergenerational assessment, thus enabling the follow-up of grandchildren bearing an edited genome.

Individuals born following HHGE should be valued in the same way as any other person, and they and their parents should not be stigmatized, or discriminated against, for having undergone HHGE.

Long-Term Monitoring

Because the consequences of genome editing for children's physical and psychological development are unknown, in order to establish whether HHGE prevents the transmission of a genetic disorder and whether there are unintended adverse and intergenerational effects it is important to assess the health of the developing fetus during pregnancy and the health and well-being of resulting children throughout the lifespan and into the next generation, if such individuals exist. Such intergenerational monitoring would be important for evaluating the well-being of the individuals involved rather than refining HHGE technologies, since the technologies are likely to have been refined in the interim. The section below briefly reviews monitoring that has been done with other ARTs and then turns to the consideration for HHGE.

Monitoring Children Born Through Assisted Reproductive Technologies

Regarding the follow-up of children born following HHGE, the closest parallels are studies of children born through ARTs, such as IVF and ICSI, who are genetically-related to their parents; children born through ARTs using donated eggs, sperm, or embryos and who therefore lack a genetic connection to one or both parents; and children born following PGT in combination with IVF/ICSI. Although most follow-up studies have focused on childhood outcomes, a small number of studies have followed up children born by ARTs to adulthood. For example, young men conceived by ICSI have been followed up to assess their fertility, and children conceived by gamete donation have been followed up to assess their psychological well-being, relationships with their parents, and thoughts and feelings about their method of conception.

Physical and health outcomes.   There exists a substantial body of research on the physical and health outcomes of children born through ARTs involving large, representative samples, although most of this research has focused on short-term rather than longer-term outcomes. A recent, comprehensive review, largely of children born through IVF and ICSI, focused on studies of singleton children to avoid the confounding effects of multiple births (Berntsen et al., 2019). It was concluded that children born following ARTs are at some risk of adverse short-term outcomes such as low birth weight, pre-term birth, and birth defects, although these risks were described as modest. The poorer outcomes for these children appeared to result from a combination of the parents' subfertility and specific aspects of the ART procedure, although it was difficult to disentangle the two in the absence of appropriate comparison groups.

There has been less research on the development of children born following PGT. In a cohort study of children born following PGT in Denmark (Bay et al., 2016), the level of adverse obstetric and neonatal outcomes was higher than that of spontaneous pregnancies but similar to that of pregnancies following ICSI. It appeared that the increased risk of adverse outcomes was related to the underlying parental genetic condition rather than the PGT procedure itself. Studies that followed children born after

PGT to age 2 found no differences in growth, cognitive and psychomotor development, and behavioral and health outcomes compared to ICSI and naturally conceived children. More recent studies of 5-year-olds (Heijligers et al., 2018) and 9-year-olds (Kuiper et al., 2018) produced similarly reassuring results.

Psychological outcomes.   Research on the psychological well-being of children born through ARTs dates back to the 1990s (for reviews see Golombok [2017, 2019]). These studies have shown that these children's families are generally characterized by positive parent-child relationships and well-adjusted children, irrespective of the presence or absence of a genetic connection between one or both parents and the child. In spite of a growing trend toward greater openness, many parents do not tell their donor-conceived children about their origins, mainly because of the concern that this information would jeopardize family relationships, especially the relationship between the non-genetic parent and the child. Parents who tell their children about their biological origins when they are young generally find that their fears about the potentially negative consequences of disclosure were unfounded, and there is growing evidence that early disclosure is associated with more positive outcomes for donor-conceived children and adults. Some children and adults who are aware of their donor-conception search for information about their donor and donor siblings (genetic half-siblings born from the same donor) in order to acquire a greater understanding of who they are and how they came to be.

These findings suggest that disclosure to children born through HHGE of the circumstances of their conception at an early age is likely to be beneficial for their psychological well-being and relationships with their parents, and that persons born following HHGE may be interested in what was done to them as embryos and why. It is not known whether, or what, parents tell children born following PGT about their origins, although they may tell them that they were selected as embryos following genetic testing. However, HHGE is distinct from PGT in that it produces an alteration to the genetic make-up of their children whereas PGT does not.

Long-Term Follow-Up of Children Born Following
Heritable Human Genome Editing

As with informed consent, a broad approach is described for the long-term follow-up of children born following HHGE, as the most appropriate specific assessments will be dependent upon a number of factors, including the conditions being edited and the countries in which the editing is carried out. Comprehensive long-term follow-up would include the assessment of (i) obstetric and perinatal outcomes; (ii) genetic disorders in resulting births; (iii) other health problems in children; (iv) growth, motor, and physical development; (v) cognitive and language development including developmental delay; and (vi) psychological adjustment including mental health problems. It will be necessary to achieve a balance between the need for monitoring and the need to avoid unduly burdening the children and adults concerned.

To examine children's growth, cognitive development, language development, and social and emotional development, assessments at key developmental milestones would be required, depending on the specific aspect of development under consideration. There are standardized tests designed for this purpose. If these assessments are to be used internationally, there is a need for versions of tests that have been translated into different languages, have been adapted to be culturally appropriate, and have normative data for the interpretation of the meaning of individual children's scores in the context of their own language and culture (Gregoire et al., 2008). There is growing consensus that it is important, both from an ethical perspective and to increase understanding, to include children's voices in interventions that affect them. For this reason, children affected by HHGE should be interviewed about their thoughts, feelings, and experiences, beginning in adolescence. A key issue for long-term follow-up studies of children born following initial HHGE uses is the need to minimize sample attrition in order to reduce sample bias. Challenges include the small number of children who would be born following initial uses of HHGE, the need to standardize both the genetic disorders under investigation and the genome editing undertaken, and the absence of meaningful comparison groups, all of which would restrict the reliability, validity, and generalizability of the findings.

In the United Kingdom, children conceived through MRT, the closest parallel to HHGE, are assessed from the prenatal period onward. During pregnancy, in addition to the monitoring of fetal growth and development, parents are offered amniocentesis. At birth, neonatal indicators are recorded, and infants receive routine developmental checks in the first year of life. The outcome of MRT on children born will be initially studied at 18 months (an age at which there are clear developmental milestones) and then subsequent follow-up will be conducted throughout childhood with appropriate consent from the parents and assent from the child, followed by consent in adulthood by the individual whose genome was edited (Gorman et al., 2018).

OTHER POSSIBLE USES OF
HERITABLE HUMAN GENOME EDITING

The majority of this chapter has focused on the use of HHGE by prospective parents to prevent transmission of a genetic variant that causes a single-gene (monogenic) disease in order to have a genetically-related child unaffected by that disease. This section discusses the potential use of HHGE in more complex circumstances: to prevent the transmission of polygenic diseases, to affect characteristics not associated with disease, and in the special circumstance of male infertility.

Human genetics is complex. Although the distinction between monogenic and polygenic conditions is useful, the reality of human genetics is more complicated. Despite the extraordinary developments in understanding mammalian gene function, relatively little is known about the function of the majority of genes in the human and mouse genomes. This ignorance about basic biology is a reminder of the research that is still required to ground clinical interventions firmly in genomic knowledge (Brown and Lad, 2019; Oprea et al., 2018).

Charting the landscape of human genetic variation and understanding its contribution to disease will require a systematic understanding of how genomic variation reliably predisposes to disease, both monogenic and polygenic. Many genes are pleiotropic; they play a role in multiple biochemical pathways. Genes operate in functional networks, and understanding the role that genes and their variants play in such networks is a challenge that will require powerful computational tools and large-scale datasets from genomics programs, including animal models (Cacheiro et al., 2020). Monogenic and polygenic inheritable conditions might also share underlying similarities. Rare single-gene disorders often predispose to more common diseases characterized by complex inheritance, with variants underlying monogenic disease interacting to contribute to complex disease risk (Blair et al., 2013).

Polygenic Diseases

Many common diseases have significant genetic contributions, including such conditions as type 2 diabetes, rheumatoid arthritis, heart disease, schizophrenia, many cancers, and Alzheimer's disease. Most of these diseases are polygenic, influenced by many genetic variants across many genes. These genetic variants typically have very small effects on the risk of disease (Timpson et al., 2018). Nonetheless, in some cases, particular variants in particular genes can have a comparatively large effect on the risk of a disease, although even then such variants do not completely determine who will get a disease or how serious it will be.[12] For example, individuals with a single copy of a common variant in the *APOE* gene, called *APOE4*, approximately doubles the risk of dementia in each decade after age 60 compared to individuals without this variant. Nonetheless, the absolute risk of dementia with one copy of *APOE4* is only approximately 5 percent from ages 60–69, and 15–20 percent over age 80 (Rasmussen et al., 2018).

In polygenic inheritance, individual DNA variants might each alter gene function modestly, most often by altering the production of a protein, and would not individually cause the disease phenotype. For example, a genome-wide association study in 11,260 cases of schizophrenia and 24,542 controls identified 145 novel associated loci, each contributing only a small amount to the risk for developing the disease (Paradiñas et al., 2018). A genetic risk score estimate can be generated by combining the different associated loci into a single score associated with a higher or lower risk for developing the disease. However, these scores indicate only relative risk and are not deterministic. Risks for most common (polygenic) diseases depend not only on complex genetic factors, but also on a wide range of environmental influences, such as diet and lifestyle choices, and on random events whose impact is difficult to predict.

We do not have sufficiently predictive information to contemplate the use of HHGE for intervening in the many common diseases that are associated with multiple variants and complex patterns of inheritance. Editing a gene variant associated with a complex disease is likely to have only a minor effect on the risk of developing that disease, while also potentially introducing unknown effects because of other biological roles the gene may play and other genetic networks in which it may interact. Genetic variants that alter the risk for one disease frequently have effects on the risk for other diseases, often in opposite directions. Such difficulties are compounded when the editing of multiple variants is contemplated. Valuable research on genetic and non-genetic factors associated with many human diseases continues, but this knowledge is far from a stage at which it could support the use of HHGE for the prevention or reduction of risk in the case of polygenic diseases.

Complex, Non-Disease Traits

Geneticists have also identified many genetic variants associated with personal characteristics, and genome editing has been proposed as a potential way to alter these. For example, genetic variants are associated with increased muscle strength or with an improved ability to increase strength

---

[12] Other types of muti-genic disease inheritance are possible, including digenic, in which mutations in two different genes are required for the disease (Deltas, 2018).

through training, and such traits are desirable for some people. An example is a variant in the *ACTN3* gene that encodes a protein that is part of muscle fiber and is associated with muscle strength (Ma et al., 2013). Some studies have found significant associations between a particular variant of this gene and muscle strength and function, though others have not (Pickering and Kiely, 2017). It is unlikely that this variant on its own is responsible for the level of muscle strength; therefore, the outcome of genome editing for this variant would be unpredictable.

As with polygenic diseases, we have insufficient knowledge of the biological effects of the many individual genetic variants associated with complex traits. Even if it were possible to edit all the variants associated with a particular trait, it would not be possible to predict the phenotypic outcome. If the purpose of the editing were to obtain greater-than-typical function, rather than preventing a disease, this would be a form of genetic enhancement. These types of uses of HHGE would likely be very controversial, raise many additional societal and ethical concerns, and be scientifically very premature.

Treatment of Male Infertility

The circumstance of male infertility represents a unique case for which genome editing could be contemplated. For a man whose infertility has an identified monogenic genetic cause, genome editing *in vivo* in his reproductive tissues (e.g., testes) or *in vitro* in his SSCs could offer the opportunity to restore fertility. This type of use sits at the intersection of somatic and heritable genome editing. The male patient in this circumstance exists and could provide informed consent to the procedure, and the clinical intent here is to treat a condition having a negative impact on that person's life, as with somatic genome editing. However, because the cells targeted for genome editing are reproductive, the correction of an infertility-causing mutation would produce a genetic change that is heritable. In its effect, genome editing for infertility would be a use of HHGE. Restoring fertility in a woman would be an analogous circumstance and is a theoretical possibility, but as discussed above it is likely to exist only through use of stem cell–derived gametes (IVG) and remains significantly further from potential clinical application than the ability to obtain and edit the genome of testicular SSCs.

Approximately 7 percent of men reportedly have some form of infertility (Krausz and Riera-Escamilla, 2018). This condition can arise from multiple potential causes, and the origin of many cases of infertility remains unexplained, but it has been estimated that more than 2,000 genes have transcripts specific to male germline cells (Schultz et al., 2003). Research continues to identify genes affecting fertility, but a number of single gene mutations are associated with quantitative and qualitative spermatogenic anomalies (Ben Khelifa et al., 2011; Harbuz et al., 2011; Kasak et al., 2018; Maor-Sagie et al., 2015; Nsota Mbango et al., 2019; Okutman et al., 2015; Tenenbaum-Rakover et al., 2015; Yatsenko et al., 2015).

For men with genetic conditions in which sperm are produced but fertility is diminished (e.g., because the sperm have reduced motility), existing reproductive options such as ICSI may be able to overcome the fertility challenge and result in the creation of an embryo. On the other hand, some men possess identified genetic mutations that make them unable to produce any sperm (azoospermia) or produce sperm with significant genetic or structural abnormalities incompatible with creating a viable embryo. For this subset of people, genome editing of SSCs and these cells' subsequent development into sperm could provide the ability to create a genetically-related child. This process may still require the use of ARTs, for example, if the SSCs are developed into sperm in culture. Or, if the genome-edited SSCs are reintroduced to the testes to develop into sperm *in vivo*, this intervention could provide an ability to have a genetically-related child without the use of IVF. Although infertility resulting from single gene mutations affects only a fraction of the human population, those who have identifiable genetic causes for this condition form a potential community that may be interested in access to HHGE.

CONCLUSIONS AND RECOMMENDATIONS

This chapter explored in detail the state of the science in genetics, genome editing technologies, and reproductive medicine that would be required for potential translational development of HHGE. It also explored circumstances in which HHGE has been proposed as a technology that could increase reproductive options. The chapter then discussed the current evidence for, and state of understanding of, the feasibility of these options. The key messages arising from this analysis are provided below.

Knowledge of Human Genetics Limits
Potential Applications of Heritable Human Genome Editing

For monogenic diseases, the use of HHGE to change a genetic variant that causes the disease to a non-pathogenic DNA sequence that is common in the population could prevent the disease being transmitted to offspring, if it is possible to efficiently and reliably make precise genomic changes without undesired changes affecting human embryos. Many human diseases are polygenic, in which a phenotype is the result of many genetic and external factors that individually have small effects and that interact in complex ways. Current knowledge is not sufficient to use HHGE to edit variants associated with the risk of developing complex, polygenic diseases or that affect non-disease traits.

Heritable Human Genome Editing Could Provide a Reproductive Option for Some Prospective Parents at Risk of Transmitting a Disease Genotype

In certain circumstances, HHGE would represent the sole option for a couple to have a genetically-related child who did not inherit a disease-causing genotype. These circumstances would arise when one prospective parent is homozygous for a dominant disease-causing mutation, or both prospective parents are homozygous or compound heterozygous for mutations in the same gene whose mutation causes a recessive disease.

In all other circumstances, at least some of a couple's embryos are expected to not have a disease-causing genotype, and PGT could offer a potential solution. Demand for PGT screening for disease-causing mutations is increasing. However, PGT is not without physical and financial costs, and some couples do not achieve a live birth after one or more cycles. Genome editing might increase the number of embryos available to a couple for transfer and thus may increase PGT success rates, although improved data concerning current PGT usage and failure rates are needed to provide insight into the extent to which there are unmet needs in this area and thus potential future demand for HHGE.

Current Technologies for Heritable Human Genome Editing Are Inadequate

Genome editing of zygotes (one-cell embryos) produced by IVF is currently the most feasible approach for carrying out HHGE, but the efficiency and specificity of current embryo-editing methods are not adequate for human clinical uses. There are no procedures to adequately control the outcome of DNA repair in embryos after an editing-induced break is introduced into the genome, and current analytical methods are not sufficient to provide necessary assessment of on-target and off-target events and mosaicism in a clinical context. Genome editing methods that avoid production of a double-strand DNA break are also subject to most of these same deficiencies. Research to refine methods and understand the feasibility and limitations of genome editing in zygotes is needed. This research would necessarily involve the use of human embryos because there are differences in the DNA repair mechanisms used by different cell types and there are important species-specific differences in early development which limits the usefulness of research in model organisms and other human cell types.

Genome editing conducted in stem cells that are induced to form male, and potentially female, gametes provides a theoretical alternative approach for HHGE. Such cells can be maintained and characterized in cell culture, but these approaches raise distinct technical, ethical, and social considerations of their own and would have to be permitted as a means of assisted reproduction in a given regulatory framework before they could be considered for use in HHGE.

Plans for Consent and Monitoring Are Needed

Any clinical use of HHGE would need to include detailed plans for obtaining informed consent and for conducting long-term monitoring of the effects of genome-editing methodologies. Established protocols and legally recognized approaches exist for obtaining informed consent from parents to participate in a clinical trial of a reproductive intervention and for obtaining consent from parents, and eventually children, to participate in future stages of monitoring and assessment. Similarly, protocols exist for the long-term monitoring and evaluation of the physical and psychological development of people born using novel ARTs. HHGE could draw on these protocols, although details would need to be adapted to the context of clinical use, and HHGE raises special considerations due to the nature of the intervention and the heritability of the genetic changes that could be made.

Gaps in Current Scientific and Technical Understanding Need to Be Addressed

Genome-editing tools are effective at making targeted DNA sequence modifications and could, in principle, be applied to HHGE. However, scientific and technical knowledge gaps would need to be addressed before any clinical use of HHGE could responsibly be considered. These gaps include the following.

Limitations in the Understanding of Human Genetics

Multiple genes influence the disease risk for the majority of human diseases and disorders, and many genetic variants identified in humans have an unknown impact on phenotype. Existing knowledge is not sufficient to predict the effects of making genetic changes in these circumstances. Even for monogenic diseases, sound evidence for a causative role of an identified genetic variant would be needed prior to genome editing. The impact of genetic background on the function of particular disease-related variants is also not well understood, and in some cases can modify the risk or clinical severity of a disease, which complicates the analysis of the potential risks and benefits of HHGE.

Limitations in the Understanding of Genome-Editing Technologies

It remains unknown to what extent the results of genome editing in model systems, such as human somatic cells and embryonic stem cells, and in other animals is predictive of the efficiency or effects of editing to correct a specific mutation in human embryos. In addition, controlling the DNA repair processes in embryos is difficult because the pathway is often dominated by complex indels produced by NHEJ. Newer variants of Cas-gRNA systems and methods such as prime editing have not yet been extensively tested in embryos.

Limitations Associated with Characterizing the
Effects of Genome Editing in Human Embryos

Understanding of the effects of unintended genetic changes, both on- and off-target, on subsequent development and long-term health remains limited. Although genome editing is commonly used to create mice and other organisms that appear developmentally normal, protocols suitable for preclinical validation of human editing would need to be established. These would need to determine (a) the efficiency of achieving desired on-target edits; (b) the frequency with which undesired edits are made, including absolute and relative frequencies of NHEJ, HDR, chromosomal translocations, and large genomic deletions or duplications resulting from editing; and (c) the frequency with which mosaic embryos arise. This will likely require developing an improved understanding of how DNA repair processes operate in germline cells, zygotes, and early embryos; how outcomes depend on the timing of introducing the editing reagents; and what formats of template DNA and delivery are most effective.

Limitations Associated with the Development of
Genome Editing in Stem Cell–Derived Gametes

Studies in animal models reveal that the use of genome editing in gamete precursor cells and the production of male gametes able to produce healthy embryos without a disease-causing genotype are a nearer-term prospect than the production of female gametes. The production of gametes from stem cells *in vitro*, if successful, would offer an alternative to HHGE in most envisaged cases. But such technology requires the same careful scientific and societal assessment needed for HHGE.

Recommendation 1: No attempt to establish a pregnancy with a human embryo that has undergone genome editing should proceed unless and until it has been clearly established that it is possible to efficiently and reliably make precise genomic changes without undesired changes in human embryos. These criteria have not yet been met, and further research and review would be necessary to meet them.

Beyond the technical considerations, countries contemplating allowing the use of HHGE will need to address the broader issues raised by the technology. As discussed in greater detail in Chapter 5, this should include engagement with a wide range of perspectives within their jurisdictions.

Recommendation 2: Extensive societal dialogue should be undertaken before a country makes a decision on whether to permit clinical use of heritable human genome editing (HHGE). The clinical use of HHGE raises not only scientific and medical considerations but also societal and ethical issues that were beyond the Commission's charge.

3

Potential Applications of Heritable Human Genome Editing

Chapter 2 described what is currently achievable through assisted reproductive technologies (ARTs) and the state of the science relevant to genome editing to prevent the transmission of a heritable disease. Chapter 2 concluded that there are significant gaps in scientific knowledge that would need to be filled before heritable human genome editing (HHGE) could responsibly be considered for clinical use. Chapter 3 addresses questions raised by defining a responsible clinical translational pathway for evaluating a potential use of HHGE, in the event that a country chooses to do so. After laying out general considerations related to the potential harms, benefits, and uncertainties of HHGE, Chapter 3 outlines six broad categories of uses for which HHGE could be considered and describes the genetic and clinical considerations associated with each category. It then sets out the Commission's conclusions about the circumstances in which a translational pathway could responsibly be described for HHGE and explains why it is not possible at present to describe a responsible translational pathway for other types of potential use.

DEFINING APPROPRIATE USES OF HERITABLE HUMAN GENOME EDITING

Decisions about the clinical use of HHGE involve issues that are complex and arguably unprecedented, because the potential range of modifications that could be made to the human genome is vast (from correcting a known disease-causing mutation to inserting new genes or regulatory elements); the effects are potentially multigenerational; and the potential scope of purposes that might someday be considered is wide-ranging, from helping prospective parents who currently have no prospect of having a genetically-related child unaffected by a severe genetic disease to, in the extreme, engaging in a program of eugenic modification of the human species. Moreover, the range of people who might experience potential benefits and harms is extensive, including prospective parents, their offspring, future generations who inherit these genetic modifications, and the society at large.

The clinical use of HHGE requires addressing two distinct issues: (1) whether a country decides that the clinical use of HHGE is appropriate for any purpose and, if so, for which purposes; and (2) if a purpose is deemed appropriate, how to define a responsible translational pathway for evaluating it with respect to efficacy and safety.

The first issue involves societal values, including ethical, cultural, legal, and religious considerations, and should be informed by scientific knowledge. Decisions about whether to permit clinical use of HHGE at all will involve weighing both deep *concerns* (ranging from the appropriateness of altering human DNA, regarded by some as a fundamental aspect of humanity, to the ultimate societal impact of widespread and extensive manipulation of the human genome) and deep *obligations* to ensure that humankind can benefit from scientific knowledge and medical advances. Decisions about the appropriateness of specific uses may depend on the extent to which they are judged to address a compelling need. Interest in using HHGE as an ART—to assist at-risk couples to have a genetically-related child that does not inherit a serious genetic disease—is driven by the recognition that many couples have a strong preference to have children who are genetically-related to both parents (Hendriks et al., 2017; Rulli, 2014; Segers et al., 2019). In contrast, interest in using HHGE to "enhance" the human species involves different motivations and raises serious issues associated with discredited projects of eugenics. Consideration of HHGE will thus need to involve careful societal decision making about whether and when to cross thresholds, informed by scientific knowledge but relying on value judgments. While very important, such considerations lie beyond the remit of this Commission.

The second issue is at the heart of the Commission's task: defining responsible translational pathways for particular uses of HHGE, should a country judge the uses appropriate. Defining responsible translational pathways clearly involves scientific considerations, but it also entails societal and ethical considerations related to weighing potential benefits and harms, and uncertainties about them, in the clinical evaluation of a new medical technology. Notably, the Commission's Statement of Task requires considering both the research and clinical issues and the "societal and ethical issues, where inextricably linked to research and clinical practice."

Below, the Commission considers the circumstances for which it concludes it is currently possible to define a responsible translational pathway only for initial clinical uses of HHGE, should a country choose to permit them. Decisions to go beyond these initial uses would depend on scientific conclusions based on experiences gained from the initial uses, societal decisions about the appropriateness, and the definition of responsible translational pathways.

CRITERIA FOR DEFINING RESPONSIBLE
TRANSLATIONAL PATHWAYS FOR INITIAL USES
OF HERITABLE HUMAN GENOME EDITING

A number of considerations are common to the initial human use of new biomedical technologies: the prioritization of safety, very careful selection of a small number of initial cases, emphasis on a favorable balance of risks and potential benefits, and careful review of initial results prior to additional uses. New interventions, with necessarily high degrees of uncertainty about their efficacy, tend to focus on diseases and individuals for whom there are no available alternatives, and on diseases or conditions for which mortality is high and/or morbidity is severe, thereby reflecting the most favorable balance of potential harms and benefits. When these considerations are met, participation in first-in-human uses can be offered to candidates after a process of informed consent.

These three considerations apply to initial uses of HHGE, along with two additional ethical issues that further support attention to the three considerations. First, because HHGE would be a reproductive technology, and as is the case in the use of any ART, prospective parents can provide consent, but the individual who would be created as a result of the technology cannot provide consent. Second, HHGE would create a heritable genetic alteration, which could be passed to future generations. This collection of considerations supports the criteria outlined below.

The Commission's approach was also informed by analyses undertaken by groups that evaluated the acceptability and considerations for undertaking the initial uses of mitochondrial replacement techniques (MRT) in humans (e.g., Bioethics Advisory Committee, Singapore, 2018; HFEA, 2016; NASEM, 2016; NCB, 2012). Although a technology with more limited potential scope of application than HHGE, MRT provides a useful starting point, given its parallels as a novel ART that creates heritable genetic changes with the aim of enabling parents to have a genetically-related child unaffected by a disease. A U.S. National Academies of Sciences, Engineering, and Medicine study concluded that "[i]n assessing the ethics of the balance of benefits and risks in MRT clinical investigations, minimizing the risk of harm to the child born as a result of MRT is the primary value to be considered," reaching this conclusion through "an approach that entails weighing, first and foremost, the probability of significant adverse outcomes borne by the children born as a result of MRT against the benefits accruing to families desiring children who are related to them through their [nuclear] DNA" (NASEM, 2016, p. 115–117). In addition, the United Kingdom gave permission to consider MRT for initial human uses only for circumstances in which preimplantation genetic testing (PGT) was very unlikely to enable prospective parents to have a genetically-related child without a serious mitochondrial disease, and under very strict regulatory oversight (see Chapter 1 for further detail on the regulatory oversight and licensing system for MRT in the United Kingdom).

The Commission distilled these general considerations into a set of principles to guide the development of a translational pathway for any initial use of HHGE:

Highest priority on safety. A combination of factors all support the need to assure the highest level of safety in initial uses of HHGE. As an ART, HHGE would be directed to creating a person without a specific genetic disease (as is the case with PGT), rather than treating an existing patient with a disease.[1] The level of safety considered acceptable for permitting initial clinical uses of HHGE should consequently be considerably higher than for somatic cell genome editing.

Most favorable balance of potential harms and benefits. In order to create the most favorable balance of potential harms and benefits, new interventions with substantial uncertainties ideally focus on diseases and prospective parents for whom there are no available alternatives, and on diseases or conditions for which mortality is high and/or morbidity is severe. The use of seriousness of a condition as a criterion for medical intervention is common in laws, regulations, and policy statements (Kleiderman et al., 2019; Wertz and Knoppers, 2002) and the seriousness of the disease in question is currently a central consideration for other ARTs that aim to prevent transmission of a heritable disease, such as PGT and MRT. Although there is not a uniformly established definition of what clinical

---

[1] In a clinical context, the nature and extent of harm is most commonly understood by comparing the health of an individual before and after a treatment. Understanding harms (and benefits) in the context of an ART such as HHGE involves comparison between the health of an individual born following genome editing of a given embryo with the anticipated health of the individual that would have been born if the same embryo had been transferred but without the prior use of genome editing. While recognizing that this raises some philosophical questions, the Commission felt that the word 'harm' intuitively captures the idea that individuals born following HHGE could be unintentionally and negatively impacted at some point by the editing procedure.

presentation is meant by "serious," the concept tends to reflect general ideas that the effects are severe or life threatening or entail substantial impairment. For the purposes of HHGE, the Commission provides below a definition of "serious disease."

Minimizing potential harms resulting from the intended edit. It is important that the consequences of the intended genome edit are well understood, both for the immediate offspring and for future generations who might inherit it, in order to be certain that an intended edit would not have unintended deleterious consequences (on its own, via genetic interactions with other loci, or via environmental interactions). At present, the best way to achieve this goal is for an edit to change a known pathogenic genetic variant responsible for a monogenic disease to a sequence that is common in the relevant population and known not to be disease-causing.

Minimizing potential harms resulting from unintended edits. To minimize the potential for harm, it is important to minimize the chance of unintended on-target and off-target edits, which could be passed to future generations, as well as indirect effects of the editing process that could affect embryo viability or developmental potential.

Minimizing potential harms by preventing genome editing when there is no prospect of benefit. For the vast majority of prospective parents at risk of transmitting a genetic disease, only a fraction of their offspring would inherit the disease (typically, 25–50 percent). It is crucial to ensure that individuals are not created by genome editing of zygotes or embryos that do not carry the disease-causing genotype, because such individuals would have been exposed to the risks of HHGE without offsetting benefits offered by the procedure.

Availability of alternatives to HHGE that might enable parents to have a genetically-related child unaffected by a specific disease. A key consideration is whether the prospective parents already have reasonable options for conceiving a genetically-related child who does not inherit a serious genetic disease. To maximize potential benefit and minimize potential harm, it would be appropriate to confine initial uses to prospective parents who lack viable options.

CRITERIA FOR POSSIBLE INITIAL USES OF HERITABLE HUMAN GENOME EDITING

Based on the above principles, the Commission identified four criteria that should be met by any proposed initial uses of HHGE, in the event that a country chooses to permit them. These criteria together emphasize safety with respect to the resulting individual and an acceptable balance of potential harms and potential benefits to that individual:

1. The use of HHGE is limited to serious monogenic diseases; the Commission defines a serious monogenic disease as one that causes severe morbidity or premature death.
2. The use of HHGE is limited to changing a pathogenic genetic variant known to be responsible for the serious monogenic disease to a sequence that is common in the relevant population and that is known not to be disease-causing.
3. No embryos without the disease-causing genotype will be subjected to the process of genome editing and transfer, to ensure that no individuals resulting from edited embryos were exposed to risks of HHGE without any potential benefit.
4. The use of HHGE is limited to situations in which prospective parents (i) have no option for having a genetically-related child that does not have the serious monogenic disease, because none of their embryos would be genetically unaffected in the absence of genome editing; or (ii) have extremely poor options, because the expected proportion of unaffected embryos would be unusually low, which the Commission defines as 25 percent or less, and have attempted at least one cycle of PGT without success.

The Commission concluded that a responsible translational pathway for initial uses of HHGE would need to meet *all four* of these criteria.

For the application of these four criteria in any specific proposed clinical use of HHGE, a case by case evaluation of potential risks and benefits would be required and should proceed under appropriate regulatory oversight.

CATEGORIES OF USES OF HERITABLE
HUMAN GENOME EDITING

To apply these criteria in practice, this section outlines possible categories of use of HHGE. The Commission identified six broad categories of potential uses of HHGE, which depend on the nature of the disease, its pattern of inheritance, and other criteria.[2] The specific diseases cited under each category are only intended as examples.

---

[2] It has been suggested that HHGE could be used to increase the chances of having a "savior sibling" (a child that is a suitable immunological match for an existing child requiring an organ or cell transplant, as discussed in Chapter 2). The Commission does not discuss this

Category A: Cases of Serious Monogenic Diseases in Which All Children Would Inherit the Disease Genotype

Disease: Serious monogenic disease, with high penetrance.

Genome editing: Change a well-characterized pathogenic variant to a common, non-disease-causing sequence present in the relevant population.

Circumstances: Couples for whom all children would inherit the disease-causing genotype. These circumstances include:

- Autosomal dominant disease. If one parent carries two disease-causing alleles (affected homozygote),[3] all children would inherit the disease-causing genotype.
- Autosomal recessive disease. If both parents carry two disease-causing alleles in the same gene (affected homozygotes), all children would inherit the disease-causing genotype.
- X-linked recessive diseases. If the prospective female parent carries two disease-causing alleles (affected homozygote) and the male parent carries a disease-causing allele on his only X-chromosome (affected hemizygote), all offspring would be affected.

The circumstances in this category are rare for two reasons.

From a probabilistic standpoint, the circumstances involve couples carrying more disease-causing alleles than typical. The circumstances can arise where there is an unusually high frequency of a disease-causing mutation in a population, a high prevalence in a population of couples who are close relatives (consanguinity), or a tendency of individuals with the disease to meet and reproduce (assortative mating). The prevalence of circumstances in Category A is discussed in more detail later in this chapter.

From a medical standpoint, the circumstances in this category apply only to a small minority of instances. Because the circumstances involve one or both parents being affected with disease, they arise only for those

---

possibility further since it does not satisfy two of the Commission's four criteria for initial uses: (1) Genome editing of the histocompatibility locus would not be an example of changing a known pathogenic gene sequence (and would also be technically very challenging, involving the need to edit multiple genes); and (2) the savior sibling would be effectively exposed to the risks of HHGE, but the benefit would accrue to someone else.

[3] Individuals carrying two disease-causing alleles are referred to as homozygous if the two disease-causing mutations are identical and as compound heterozygous if the two disease-causing mutations are different. Aside from the discussion in Chapter 2 of the added complexity of editing multiple alleles in compound heterozygotes, this distinction does not matter for the purposes of this chapter, and the term homozygous will be used throughout.

serious monogenic diseases that are compatible with individuals surviving to reproductive age with preserved fertility.

Examples of serious monogenic diseases for which the circumstances in this category can arise include autosomal dominant diseases such as Huntington's disease; and autosomal recessive diseases such as cystic fibrosis (CF), sickle cell anemia, and beta-thalassemia.

Considerations: This category is unique in two important respects. First, for couples in this category prenatal diagnosis and PGT, which can identify fetuses and embryos that have not inherited the disease-causing genotype, have no chance of identifying genetically unaffected embryos. Second, the embryos exposed to risks associated with genome editing procedures would be only those carrying the disease-causing genotype for a serious monogenic disease. This stands in contrast to the situation in Category B, below.

Category B: Serious Monogenic Diseases in Which Some, but Not All, of a Couple's Children Would Inherit the Disease-Causing Genotype

Disease: Serious monogenic disease, with high penetrance.

Genome editing: Change a well-characterized pathogenic variant to a common, non-disease-causing DNA sequence present in the relevant population.

Circumstances: Couples for whom some children would inherit the disease-causing genotype. The typical circumstances are:

- Autosomal dominant disease. If one parent carries one copy of a disease-causing allele (affected heterozygote), on average 50 percent of children would inherit the disease-causing genotype and 50 percent would not.
- Autosomal recessive disease. If both parents are unaffected heterozygous carriers for disease-causing alleles, on average 25 percent of children would inherit the disease-causing genotype and 75 percent would not.
- X-linked dominant disease. If the mother is heterozygous for the disease-causing allele, on average 50 percent of all children would inherit the disease-causing genotype.
- X-linked recessive disease. If the mother is a heterozygous carrier and the father does not carry the disease-causing allele, 50 percent of male offspring on average would inherit the disease-causing genotype. Fifty percent of female offspring will be heterozygous carriers and will typically be clinically unaffected, although exceptions some-

times occur due to skewed inactivation of the genetically unaffected X chromosome, resulting in females with varying manifestations of disease.

In rare circumstances, the expected proportion of affected offspring can be higher. If both parents are heterozygous for a disease-causing allele for an autosomal dominant disease, on average 75 percent of children would inherit the disease-causing genotype. If one parent is an affected homozygote for an autosomal recessive disease and the other is a heterozygous carrier, 50 percent of the children would be expected to inherit the disease-causing genotype. These instances are expected to occur more frequently in populations with "founder effects" in which the contemporary population is derived from a small founding population, resulting in reduced allelic diversity, with particular disease-causing alleles persisting at relatively high frequency in the population.

Couples in Category B are far more common than couples in Category A, for two reasons. From a probabilistic standpoint, Category B typically involves individuals carrying one rather than two disease-causing alleles for a given disease. From a medical standpoint, Category B comprises many more diseases than Category A. Because the parents in Category B may be unaffected carriers in the case of autosomal and X-linked recessive diseases, the category includes all of the thousands of serious recessive and X-linked diseases. In contrast, because Category A involves only affected parents, it includes only the small subset of serious diseases for which affected individuals can survive to reproductive age.

The proportion of all reproductive couples that fall in Category B is substantial. The World Health Organization estimates that a monogenic disease is present in 1 percent of global births (WHO, 2019b). Only some of these instances fall into Category B, because some monogenic diseases do not meet the Commission's definition of being serious for the purpose of defining a responsible translational pathway for HHGE, and some couples have affected children due not to inherited mutations but to newly arising (*de novo*) mutations, which by definition could not have been prospectively identified in the parents. The Commission estimates that Category B comprises at least 0.1 percent of all couples, estimated at more than 1 million couples worldwide.

While most cancers arise from a constellation of somatic mutations, there are individuals with inherited cancers due to single mutations with high penetrance, which in some cases can be prevented by surgical removal of target tissues or organs. For example, familial adenomatous polyposis has virtually 100 percent penetrance without surgical removal of the colon. In the setting of very high penetrance, such cancer syndromes could be considered serious inherited diseases. Other inherited cancer syndromes feature lower penetrance, and other considerations would need to be taken into account in an assessment of potential harms and benefits. Additionally, other inherited variants may make more modest contributions to cancer risk, as discussed below in Category C or D.

Considerations: Category B differs from Category A in two important respects.

First, the application of HHGE, as currently conceived, would involve treating all zygotes at the single-cell stage, regardless of their genotype. Category B would therefore involve subjecting all embryos to risks associated with genome-editing procedures—including those that do not have the disease genotype and thus do not require genome editing. Possible alternative approaches for the application of HHGE that would avoid editing unaffected embryos are discussed below.

Second, couples in Category B currently have existing options for having an unaffected child genetically-related to both parents—specifically, embryo selection via PGT. The limitation here is quantitative rather than qualitative. For any given couple, the probability of success in producing a child is somewhat lower for PGT than for in vitro fertilization (IVF) in general. As noted in Chapter 2, it is estimated that approximately 80 percent to 90 percent of PGT cycles in which at least one embryo reaches the diagnosis stage result in an unaffected embryo that can be transferred to the uterus. The remaining 10 percent to 20 percent of PGT cycles do not result in transfer of an unaffected embryo, and some couples may not obtain an unaffected embryo even after several PGT cycles (particularly those from whom only a small number of eggs can be harvested).

It has been proposed that, if HHGE worked with high efficiency and safety, it might assist certain couples (those who currently have a small number of unaffected embryos) by increasing the proportion of unaffected embryos available for uterine transfer. However, this outcome is by no means certain, because the additional laboratory procedures involved in HHGE might decrease the yield of high-quality embryos available for transfer.

Category C: Other Monogenic Conditions with Less Serious Impacts Than Those in Categories A and B

Disease: Monogenic disease or disability with less serious impacts than those in Categories A and B.

Genome editing: Change a well-characterized pathogenic variant to a common, non-disease- or non-disability-causing sequence present in the relevant population.

Circumstances: This category involves prospective parents for whom all or some of their naturally conceived children would inherit the genotype that causes the monogenic condition.

An example is familial hypercholesterolemia (FH), which is caused by mutations in the gene encoding the low-density lipoprotein (LDL) receptor (or mutations in a number of other genes such as *APOB* and *PCSK9*) and can occur in heterozygous or homozygous form. Heterozygous FH is a relatively common genetic condition (with a frequency of approximately 1 in 250) that causes elevated LDL cholesterol levels that predispose to early cardiovascular morbidity and mortality. Moreover, LDL levels in these individuals can usually be effectively reduced with medications, substantially lowering the risk of heart attack, reduced quality of life, and premature death. By contrast, homozygous FH, which is rare (with a frequency of around 1 in 160,000–300,000 globally, but higher in populations with FH founder variants) causes an extreme form of hypercholesterolemia that is difficult to treat and typically leads to life-shortening heart disease (Cuchel et al., 2014), although new therapeutic approaches are being developed to effectively treat these patients. The use of HHGE in a case in which both parents carry a disease-causing FH allele would fit in Category B. When only one parent is a carrier, the case would belong in Category C, since only heterozygous or unaffected embryos could result.

A second group of examples involves genotypes that may affect an individual's quality of life but are not serious monogenic diseases within the meaning of the Commission's definition (a disease that causes severe morbidity or premature death). Inherited deafness would be an example. While some deaf individuals consider deafness as severely impacting quality of life and a condition to be avoided, others strongly disagree (Padden and Humphries, 2020). The Commission recognizes that a country's consideration of genome editing for conditions such as deafness raises many complex issues that are beyond this report's scope.

Considerations: Although Categories B and C both comprise monogenic disorders, compared with Category B the conditions in Category C feature less severe morbidity, and risk of premature death may be mitigated by relatively simple medical or lifestyle interventions.

Category D: Polygenic Diseases

Disease: Polygenic diseases, for which a large number of genetic variants each contributes to disease risk, with the variants collectively having substantial—though not determinative—effect on disease occurrence or severity.

Genome editing: Changing one, several, or even large numbers of genetic variants associated with higher risk of the disease to alternative common variants that are associated with lower risk of the disease.

Circumstances: The risk of developing common diseases is influenced by many genetic variants (often hundreds or more), as well as by interactions with non-genetic factors collectively referred to as environment (which may include diet, pathogen exposure, exercise, and much more). Most of these genetic variants are common alleles that have small effects on disease risk (altering risk by less than a factor of 1.1-fold), although a few common variants can have relatively large effects on this risk and rare alleles in some genes can have large effects on risk of common disease. The combined effects of these risk variants are often additive, although there may sometimes be genetic interactions (i.e., the presence of one variant may alter the effect of another variant).

   Examples in Category D include many common diseases, such as type 2 diabetes mellitus, heart disease, and schizophrenia, although rare Mendelian forms of common diseases can occur. For common polygenic diseases, changing a single genetic variant would typically be expected to have negligible effect on the risk of disease. A notable exception is the E4 allele of the *APOE* gene: the risk of developing Alzheimer's disease rises with every decade of life after age 60, but risk increases more rapidly depending on whether an individual has zero, one, or two copies of the E4 allele (Rasmussen et al., 2018). Even so, the *APOE* gene explains only a fraction of the risk for Alzheimer's disease (absolute risk of ~5 percent from ages 60 to 69, and 15–20 percent over age 80).

Considerations: Current scientific understanding suggests that genome editing to alter one or more gene variants associated with a polygenic disease would be unlikely to prevent the condition and might have undesired effects, as the targeted alleles may play important roles in other important biological functions and may interact with the environment. Moreover, potentially better options may be available or become available to minimize the risks of developing the disease or to help manage its consequences.

Category E: Other Applications

Disease: This category does not involve heritable diseases. Rather, it involves genetic changes directed toward other objectives, which may or may not be health-related and may involve introducing genetic sequences that do not naturally, or only very rarely, occur in the human population.

Genome editing: Genetic changes ranging from single-base substitutions to introduction of new genes or disabling of existing genes.

Circumstances: A vast range of applications can be imagined for HHGE, ranging from attempts to prevent or protect against infectious diseases, to genetic changes that would enhance normal human traits, to introducing genes conferring new biological functions. All of these applications raise scientific, societal, and ethical questions that are impossible to resolve given the current state of scientific understanding. Examples include:

- attempting to provide offspring with resistance to an infectious disease by editing a gene, for example, the attempt to inactivate the *CCR5* gene and confer resistance to HIV infection;
- attempting to produce an ability in offspring by introducing a rare allele of a specific gene known or believed to be associated with a desired phenotype. For example, constitutive activation of the *EPO* gene has been proposed to confer advantages in endurance sports (Brzeziańska et al., 2014);
- attempting to modify traits such as height or cognitive ability that are influenced by hundreds or thousands of genetic variants across the genome; and
- attempting to confer new abilities, not found in humans, by adding sets of genes that, for instance, might confer resistance to radiation exposures encountered during extended spaceflight.

Considerations: In all of the cases mentioned above, the potential impacts of HHGE on children, adults, and future generations cannot be fully assessed. For example, while it is clear that homozygous loss of *CCR5* function confers partial protection from HIV infection, this loss may increase other risks of morbidity. Moreover, effective methods of preventing and treating HIV infection are available. Similarly, a lifelong increase of red blood cell mass due to constitutive expression of erythropoietin may increase endurance but might also increase the lifetime risk of thrombosis. For these reasons, the benefit-to-harm ratio in these scenarios is uncertain and in many instances may be very low.

In addition to these scientific and clinical complexities there are, of course, numerous ethical and social obstacles to interventions in this category. Any future justification for pursuing such interventions would require both scientific agreement that the long-term impact of such changes can be assessed and societal approval about the acceptability of such interventions.

Category F: Monogenic Conditions That Cause Infertility

A special category for which genome editing might be used is in treating germline cells (or their precursors) from an existing individual to reverse infertility with a monogenic cause. In this case, genome editing would change the sequence of a gene to restore fertility. Whereas HHGE in Categories A through E would not be directed at offering therapy for an existing individual suffering from a disease but rather would be a form of assisted reproduction, Category F has the unique feature that the intended beneficiary of the genetic alteration would be an existing individual (the infertile prospective parent) with the additional impact that the edited genome would be transmitted to offspring.

This category remains hypothetical for now because, leaving aside the issues of genome editing, it is not currently possible to generate functional gametes from human stem cells. Any developments in this area would require regulatory approval for a range of ARTs before clinical applications could be considered.

CIRCUMSTANCES FOR WHICH A RESPONSIBLE TRANSLATIONAL PATHWAY COULD BE DEFINED

The Commission then considered the circumstances within the categories above that could meet the four criteria outlined earlier in the chapter for which a responsible translational pathway could currently be described. Based on this analysis, the Commission concluded that initial uses of HHGE would need to be restricted to Category A and a very small subset of Category B, provided certain conditions can be met. This section discusses Categories A–F in turn.

Category A

Category A clearly meets the four criteria for initial uses of HHGE: (1) The category involves serious monogenic diseases. (2) Genome editing would be directed at changing a pathogenic variant known to be responsible for the serious monogenic disease to a sequence commonly carried in the relevant population. (3) No individuals resulting from edited embryos could have been exposed to potential harms from HHGE without potential benefit, because all of the couple's embryos carry the disease-causing genotype. (4) Couples currently have no other options to produce a genetically-related child free of the disease.

Category B

Category B would not, as a whole, be suitable for initial uses of HHGE, because it does not currently meet the third criterion and because most couples would not meet the fourth criterion. The key difference from Category A is that couples in Category B can produce children who do not inherit the disease-causing genotype (in typical cases, at least half on average). With respect to the third criterion, HHGE, as currently conceived, would involve subjecting all zygotes (both those that do and do not have the disease-causing genotype) to genome editing procedures and thus would result in the birth of children derived from embryos that had been needlessly exposed to potential harms from genome editing. With respect to the fourth criterion, the vast majority of couples already have a viable option (PGT) for producing a genetically-related child that is free from the genetic disease. As discussed above, the substantial majority (80–90 percent) of PGT cycles in which at least one embryo reaches the diagnosis stage results in an unaffected embryo that can be transferred to the uterus. The primary interest in HHGE in Category B is to assist couples who have very low prospects of having an unaffected child, owing to few unaffected embryos being available for transfer.

After extensive discussion, the Commission concluded that initial uses of HHGE might be appropriate under *certain* circumstances for a very small subset of Category B.

First, reliable methods would need to be developed that ensure that no individual would be produced from embryos that had been needlessly subjected to HHGE, ideally by identifying embryos that carry the disease-causing genotype before performing HHGE. One approach might be to use polar-body genotyping, which has the potential to identify zygotes that have inherited from the mother an allele that causes a dominant monogenic disease (see Chapter 2); the reliability of polar-body genotyping for this purpose would need to be established. Those zygotes could be subjected to HHGE followed by PGT, while the other zygotes could be subjected to standard PGT. Another approach might be to develop reliable procedures to perform HHGE on multicellular embryos without producing embryos that are mosaic for the edit. This approach would enable the genotype of an embryo to be determined prior to delivering editing reagents. However, no such procedures are currently available.[4]

---

[4] In theory, a third approach would be to perform HHGE on all zygotes, subsequently identify by PGT those embryos that did not have the disease-causing genotype prior to genome editing and had therefore been needlessly subjected to HHGE (because HHGE targets the disease-causing mutation(s), this would require genotyping a sufficient set of polymorphic sites on each side of the mutation to distinguish the two haplotypes in each parent), and ensuring that those embryos are not transferred. However, many Commission members viewed this approach as problematic because it would require a commitment to discarding embryos that

Second, initial uses would need to be restricted to those couples that have very poor prospects for having an unaffected child with conventional PGT. The Commission defines such couples as those (i) for whom the expected proportion of unaffected offspring is 25 percent or less (for example, couples in which both parents are heterozygous for the same or different dominant serious monogenic diseases) and (ii) who have undergone at least one cycle of PGT without success, since many couples will produce enough embryos to yield unaffected embryos suitable for transfer without editing.

To meet all four criteria, any initial uses of HHGE in Category B should be confined to these circumstances.

Categories C Through F

Category C involves genetic diseases that have less serious effects, may be manageable using other methods, and may not be seen as negatively impacting quality of life by members of communities affected by the condition. Until much more is known about the safety and efficacy of HHGE, it is unclear that the potential benefits outweigh the potential harms. A cautious approach argues against undertaking first-in-human uses in this category.

Category D (polygenic diseases) and Category E (genetic changes that are not directed toward variants involved in heritable diseases and may involve genetic sequences that do not naturally occur in human populations and uses that could be seen as enhancements) are not currently suitable for HHGE. Scientific understanding and existing technologies are insufficient to produce predictable, well-characterized results, including across a range of genetic and environmental interactions, and to minimize the effects of unknown and speculative risk. Moreover, these uses raise additional societal and ethical concerns.

Category F (monogenic conditions that cause infertility) remains speculative at present, making it impossible to define a responsible translational pathway. Since human stem cell–derived in vitro gametogenesis has not been developed or permitted anywhere for medical use, it is premature to consider how it might be used in combination with HHGE.

---

would have been suitable for transfer but for the fact that they had been needlessly subjected to potential harms from HHGE.

Circumstances for a Responsible Translational Pathway for Initial Uses of Heritable Human Genome Editing

In summary, the Commission concluded that a responsible translational path for initial uses of HHGE

(1) could be defined for Category A;

(2) might be defined for the very small subset of couples in Category B who have a very low likelihood of success through PGT due to genetic circumstances (embryos having a 25 percent or lower probability of not inheriting the disease-causing genotype) and who have attempted at least one PGT cycle without success, provided that reliable methods are established to ensure that no individuals result from embryos that were needlessly subjected to HHGE; and

(3) cannot currently be defined for the rest of Category B or for Categories C through F.

As previously discussed, prior to any clinical use of HHGE in any circumstances, it will be necessary to demonstrate a safe and effective methodology and for any country offering HHGE to have an appropriate regulatory framework to oversee it. Before crossing the threshold of undertaking clinical uses of HHGE in other circumstances beyond those described above, an appropriately constituted international body should assess whether and under what circumstances a responsible translational path can be defined.

HOW COMMON ARE THE CIRCUMSTANCES FOR THE INITIAL CLINICAL USES OF HERITABLE HUMAN GENOME EDITING?

The Commission next considered the frequency of the circumstances for initial uses of HHGE defined above, to determine whether there is likely to be an adequate number of suitable couples to enable initial studies to evaluate efficacy and safety, which we judge to be approximately 10–20 couples. Our analysis suggests that there is likely to be an adequate number of prospective parents to reach this goal.[5]

As discussed below, prospective parents who might be offered HHGE would likely come from multiple countries. This observation reinforces

---

[5] These initial studies would be evaluated for the safety and efficacy of the editing and the likelihood of a successful pregnancy. This would give crucial information for further studies and it would be essential that information about these outcomes be shared. If these studies did not raise concerns about the safety or efficacy of the HHGE technique then much larger studies would be required to evaluate long-term outcomes for the individuals whose genomes had been edited.

the value of global coordination of any clinical use of HHGE. It would be important to use a clear mechanism, such as an international consortium, to identify potential participants, undertake the genome editing intervention according to the translational pathway described in this report, and evaluate clinical outcomes. Precedents exist for such international coordination and collaboration, such as the International Rare Diseases Research Consortium (Lochmüller et al., 2017), including global coordination of clinical trials.

How Common Are the Circumstances in Category A Expected to Be?

The circumstances in Category A are very rare. This is appropriate for the initial use of a technology such as HHGE, where it would be suitably cautious to begin with a small number of couples who have no alternatives, proceed carefully, and intensively study the results. It is important to assess whether there is a sufficient number of couples in Category A that could potentially benefit from HHGE. As noted above, Category A arises only for the minority of serious monogenic diseases that are compatible with individuals surviving to reproductive age and being able to reproduce. Examples of diseases where this is the case are Huntington's disease, CF, sickle cell anemia, and beta-thalassemia.

The actual number of couples in Category A is not known, although there are anecdotal examples. Basic principles of population genetics can provide an initial insight into the expected frequency of couples in Category A. Under the classic assumption of a closed, randomly mating population (specifically, individuals choose partners from within the population, their choice is not correlated with relatedness or disease status, and disease status does not affect fertility), the expected proportion of couples in Category A will be approximately $2q^2$ for an autosomal dominant disease and q^4 for an autosomal recessive disease, where q is the frequency of disease-causing alleles.[6]

The frequency of disease-causing alleles differs among diseases, depending on the rate of appearance of new mutations that give rise to new disease-causing alleles and the rate of their removal from the population via natural selection. Alleles that cause serious dominant diseases are typically much rarer than alleles that cause recessive diseases because the latter are only subjected to negative selection when an individual contains disease-causing alleles on both chromosome copies, while dominant alleles are virtually always under negative selection because only one mutant copy is

---

[6] The frequency of homozygotes is q^2. For an autosomal recessive disease, both parents must be homozygous ($q^2 \times q^2$). For an autosomal dominant disease, either parent in a Category A couple may be homozygous (approximately $q^2 + q^2$).

needed to produce disease. The collective frequency of all disease-causing alleles in a gene (q) is often in the range of 4.5×10^{-3} for a serious autosomal recessive disease and 2×10^{-5} for a serious autosomal dominant disease.[7] From these values, the expected frequency of couples in Category A occurring by chance for a particular gene would be expected to be in the range of 4×10^{-10} for a recessive disease and 8×10^{-10} for a dominant disease — that is, in the range of 4–8 per 10 billion for any given disease gene. If there were 100 similar genes in this category, the total frequency of couples in Category A would be about 100-fold higher (about 4–8 per 100 million couples). Applying similar reasoning, a recent article estimated that there would be only a small number of births from Category A circumstances in the U.S. population (Viotti et al., 2019).

The actual frequency of couples in Category A is expected to be significantly higher in populations that have much higher frequencies of certain disease alleles. For recessive monogenic diseases, an allele frequency of around 3 to 10 percent would correspond to couples in Category A occurring at frequencies between 1 in 10,000 and 1 in 1.2 million. In populations with high rates of consanguineous unions (couples who are closely related genetically) and with local variation in allele frequency, the frequency of homozygotes will be higher and therefore so will the expected frequency of couples in Category A. For dominant monogenic diseases, an allele frequency of 0.1 to 1 percent would correspond to a homozygote frequency between 1 in 10,000 to 1 in 1 million.

On the other hand, when estimating the frequency of couples in Category A, one must take into account the fact that some serious monogenic diseases shorten lifespan or decrease fertility, and some autosomal dominant diseases have more severe disease manifestations in homozygotes than heterozygotes (Homfray and Farndon, 2015; Zlotogora, 1997).

Beyond estimating disease-allele frequencies, another consideration is that, for certain recessive diseases, heterozygous carriers enjoy a benefit in certain environments. This is the case for sickle cell disease (SCD) in areas where malaria is prevalent. In such areas, people with one sickle cell allele who contract malaria are less likely to die from the disease (Archer et al., 2018).

For any disease, it is important to consider whether it is technically feasible to reliably edit the disease-causing mutation. Huntington's disease, for example, is caused by an expanded number of trinucleotide repeats within

---

[7] Under mutation-selection balance in a randomly mating population, the equilibrium frequency q is expected to be $(\mu/s)^{1/2}$ for a recessive disease and μ/s for a dominant disease, where μ is the mutation rate of new disease-causing alleles and s is the selection coefficient against the affected genotype. The values of μ and s depend on the gene. The figures cited in the text correspond to a mutation rate of new loss-of-function alleles of $\mu = 10^{-5}$ for a 'typical' human gene and a selection coefficient $s = 1/2$.

the gene. HHGE would require reducing the number of these repeats to a non-disease-causing level, which is technically more difficult than changing a single nucleotide. Alternatively, genome editing could be used to introduce sequences that do not naturally occur in the human population (e.g., a stop codon to inactivate the gene); however, our second criterion (above) restricts initial uses of HHGE to producing naturally occurring alleles that are common in the relevant population.

The examples of Huntington's and sickle cell anemia demonstrate that even among serious monogenic diseases where affected individuals survive to an age when they could have children, there are genetic and environmental factors that complicate the analysis of potential harms and benefits arising from HHGE.

Potential Examples of the Circumstances in Category A

Actual data are not readily available from the literature on numbers of couples of reproductive age in Category A. Nevertheless, as noted in the section above, very approximate estimates may be generated under simplified assumptions of random mating. The estimates suggest that couples in Category A for recessive monogenic diseases may occur at meaningful frequencies in populations with disease-allele frequencies exceeding about 3 percent, with the frequency being even higher in populations with higher rates of consanguinity. Moreover, the number of couples in Category A for dominant monogenic diseases will depend on the frequency of individuals who are homozygous for the disease-causing alleles and can and wish to have children. The following are some examples where there may be a substantial number of couples in Category A.

Beta-Thalassemia in Global Populations

Beta-thalassemia is an autosomal recessive blood disorder that disrupts the formation of hemoglobin and can cause severe anemia and other issues. Patients who produce no functional beta globin (beta-thalassemia major) require regular blood transfusions; those who produce beta globin with significantly reduced function can exhibit a range of disease severity. Without access to regular treatment, thalassemia major patients may die in adolescence, but with improved medical care, life expectancy has risen into the 40s and 50s. Mutations that cause thalassemias are relatively common, with approximately 1.5 percent of the global population estimated to be heterozygous carriers for beta-thalassemia (i.e., up to 80 million people), with high carrier rates noted across the Mediterranean region, Middle East, India, Southeast Asia, and Pacific Islands (De Sanctis et al., 2017). For example, it has been estimated that 4.5 percent of the Malaysia population are carriers for beta-thalassemia (George, 2001), indicating an allele frequency of 2.25 percent. In a population of roughly 32 million this suggests approximately 10 homozygous couples in that country. In India, the carrier rate is estimated to be between 3 to 4 percent (GUaRDIAN Consortium et al., 2019), indicating allele frequencies of 1.5 to 2 percent. In a population of roughly 1.35 billion people, this suggests there may be between approximately 70 and 200 homozygous couples for beta-thalassemia. In North African countries, estimates of beta-thalassemia carrier rates range from 1 to 9 percent, suggesting allele frequencies in the range of 0.5 to 4.5 percent (Romdhane et al., 2019). Considering just the North African population of about 240 million, the expected frequency of homozygotes is sufficiently high (approximately 1 in 500 to 1 in 40,000, depending on the region) that there may be many couples with both members homozygous for beta-thalassemia.

Sickle Cell Disease in Sub-Saharan Africa and the United States

SCD is an autosomal recessive disorder occurring when an affected individual carries two copies of the allele for sickle cell trait. The prevalence of sickle cell trait is high in many populations in sub-Saharan Africa, due to the heterozygote advantage described above. In one example, screening of several thousand women of child-bearing age and their male partners in the Enugu state of Nigeria (population 3.3 million)[8] identified sickle cell trait in 22 percent of individuals (Burnham-Marusich et al., 2016). Based on this frequency, the authors expected to identify approximately 1 percent of their study cohort as having SCD but identified only 0.1 percent of their study cohort as being SCD homozygotes; they speculated that this may be due to early mortality, which has been estimated to be 50 to 90 percent for SCD in sub-Saharan Africa. This rate of reproductive-age SCD homozygotes would suggest a frequency of couples in Category A of approximately 1 per 1 million in this population. The situation is expected to be similar in the many other areas of sub-Saharan Africa in which sickle cell trait is common, suggesting that there could be hundreds to potentially thousands of homozygous couples across those areas where SCD is most prevalent. The frequency of sickle cell trait in the African American population is also relatively high (estimated at roughly 7 percent),[9] with more than 90 percent of SCD homozygotes estimated to live past age 18 and commonly into their 40s (Platt et al., 1994; Quinn et al., 2010). Viotti et al. (2019) used this carrier frequency to estimate that there are approximately 80 homozygous couples among African Americans.

---

[8] See https://www.enugustate.gov.ng.
[9] See https://www.cdc.gov/ncbddd/sicklecell/data.html.

Cystic Fibrosis

The carrier frequency for a mutation in the gene that causes CF, an autosomal recessive disease, is approximately 1 in 30 (around 3 percent) in Caucasian Americans (Strom et al., 2011), resulting in CF in approximately 1 in 3,600 births. Similar estimates for incidence of CF are reported for European populations (Farrell, 2008). The authors of a recent paper estimated that there are only 1–2 reproductive-age couples in the United States in which both parents are homozygous for CF (Viotti et al., 2019). Based on similar CF allele frequencies and the roughly 1.5 times greater population, one could expect several such couples in Europe. The rapid advances in the treatment of CF may result in an increased number of couples in which both people are affected by CF being able to have children.

How Common Are the Circumstances in the
Subset of Category B Expected to Be?

To fit the circumstances of the very small subset of Category B, both prospective parents would need to be heterozygous for the same or different serious dominant disease(s). Such circumstances are expected to be rare as they depend on both parents carrying disease-causing alleles and on people with the disease surviving to reproductive age and being able to have children. Some examples of diseases that might be compatible with these circumstances are Huntington's disease, early onset Alzheimer's disease, and familial adenomatous polyposis.

Huntington's is a neurodegenerative disease that arises from an expanded number of three nucleotide repeats in the DNA sequence of the gene *HTT*. The disease is found in approximately 3–7 per 100,000 people of European descent[10] and has been estimated to be 12.3 per 100,000 people in the United Kingdom (Evans et al., 2013). Random assortment of couples would lead to roughly 1 couple per 67 million couples in which both parents are heterozygous carriers, corresponding to roughly 3 couples in the United States and Europe combined. As noted earlier in the chapter, to meet the criteria identified by the Commission for initial clinical uses of HHGE would also require having a genome editing methodology capable of reducing the number of trinucleotide repeats to a level typical of unaffected individuals.

Mutations in the gene presenilin 1 (*PSEN1*) cause early onset Alzheimer's disease. Although *PSEN1* mutations are the most common cause of early onset inherited Alzheimer's disease, determining the frequency of *PSEN1* mutations in a population is complicated by the fact

---

[10] See https://ghr.nlm.nih.gov/condition/huntington-disease#statistics.

that multiple possible mutations (not only in *PSEN1* but also in the genes *PSEN2* or *APP*) can cause this disease and by the fact that there are also later onset forms of Alzheimer's and dementia. It has been estimated that up to 1 percent of Alzheimer's cases arise from gene mutations in *PSEN1*, *PSEN2*, and *APP*;[11] other estimates have indicated that 50,000 to 250,000 people in the United States have early onset Alzheimer's disease, occurring prior to age 65.[12] Estimates of the frequency of *PSEN1* mutations in global populations are not readily available. Cases in which both prospective parents carry the mutation may be more common where there are higher rates of consanguineous marriage.

Mutations in the gene *APC* cause the disease familial adenomatous polyposis, which results in the development of colon cancer by middle age as well as increased risk of cancer in other organs. Familial adenomatous polyposis has been reported to occur in 1 in 7,000 to 1 in 22,000 people.[13]

Although both parents would need to carry alleles for a serious dominant disease to meet the circumstances identified by the Commission for potential initial uses of HHGE, it may not be necessary for parents to carry alleles for the same disease. It could be possible that each parent is heterozygous for a different dominant disease. By probability, the embryos such a couple could produce would still have only a 25 percent chance of being unaffected by a serious disease. However, the use of HHGE in such a circumstance would entail decisions about whether to attempt genome editing of more than one disease-causing allele or which disease to target through the editing process.

These examples help illustrate that circumstances in this very small subset of Category B are likely to be rare. However, such cases are expected to exist. Dominant mutations may be found at significant frequency in founder populations, and union between individuals with the same or different mutations is not rare. In addition, several PGT clinics in the United States and Western Europe indicated that they have seen patients whose embryos would have a low chance of being unaffected by a genetic disease (personal communications). Although detailed data were not available to the Commission, a preliminary estimate was up to 1 such couple per year per clinic, as compared to 50–100 couples seen for circumstances in Category B where embryos would have a 50 percent chance of inheriting a disease-causing genotype.

---

[11] See https://www.alz.org/alzheimers-dementia/what-is-alzheimers/causes-and-risk-factors/genetics.

[12] See https://www.alzforum.org/early-onset-familial-ad/overview/what-early-onset-familial-alzheimer-disease-efad.

[13] See https://ghr.nlm.nih.gov/condition/familial-adenomatous-polyposis#statistics.

Considerations After Initial Human Uses

Should first-in-human uses take place and appear to be successful, without raising concerns about safety and efficacy, it may become appropriate to consider the use of HHGE in additional circumstances in Category B. Such a decision could enable evidence to be obtained on whether or not HHGE followed by PGT provides an improved option compared to PGT alone for prospective parents wishing to prevent transmission of a serious monogenic disease. However, this would require that a controlled clinical evaluation (randomized control trial) be designed to compare the success rates of these two types of interventions (PGT alone in one arm versus HHGE with PGT in the other). Such evidence would answer questions that have been raised about whether, in particular genetic settings, HHGE can increase the numbers of high-quality embryos available for transfer for couples in which some embryos will inherit disease-causing genotypes, and the results would inform discussions on future clinical practice. The numbers of couples who would take part in any initial clinical uses of HHGE to evaluate safety and efficacy is expected to be too small to design and recruit participants for such an evaluation. Moreover, the comparison would depend on the genetic setting (specifically, the expected proportion of unaffected embryos). Conducting such evaluation would thus require the inclusion of many additional participants in Category B. Evaluating the results of any initial human uses and making decisions on whether to consider any further uses of HHGE would require national and international processes described in Chapter 5.

THE NEED FOR CONTINUED RESEARCH

Fundamental laboratory research (not undertaken with the clinical aim of establishing a pregnancy) related to genome editing of human gametes, zygotes, and embryos is itself important.

To Better Understand Human Embryo Development

The understanding of human embryo development is an important area of research. Genome editing has already provided major new insights into preimplantation human development. Such research on human embryos, while raising ethical issues of great importance, is scientifically essential because there are considerable differences between species. Such studies will lead to better understanding of the reasons for the limited success of IVF for some prospective parents and may well help our understanding of female infertility and miscarriage. Research using genome editing in human embryos will also give important insight into the effects of maternal aging on human embryo development, an area of increasing interest with a growing number of women choosing to delay pregnancy. It will also shed light on mechanisms of DNA repair that operate specifically in the early embryo, a process that will inevitably need to be controlled in order for the outcomes of genome editing to be completely predictable and precise. Finally, it will help in understanding the role that key genes play in specifying cell fate in the human embryo, which may have profound implications for our ability to culture and manipulate human stem cells for applications in regenerative medicine. To perform such research to the highest standards, in which a particular embryonic phenotype can be attributed to a specific genetic event, researchers will require genome editing protocols of the highest efficiency and specificity.

To Improve on Assisted Reproductive Technologies

Fundamental research to improve the general ability to precisely edit the human genome, control on-target events, avoid mosaicism, and generate no off-target effects could improve the utility of HHGE in an assisted reproduction context. If HHGE could be performed very safely and at extremely high efficiency, it could be possible to use it to increase the number of embryos not carrying the disease-causing genotype available to prospective parents undergoing PGT, which might allow expanding use broadly in Category B.

CONCLUSIONS AND RECOMMENDATIONS

It is not possible to define a responsible translation pathway for all possible uses of HHGE, because the benefits and risks depend on particular circumstances, including the severity of the disease, the genetic situation of the couple, the mode of inheritance of the disease, the nature of the proposed sequence change, and the availability of alternatives. Given the uncertainties inherent in a new technology like HHGE, clinical evaluation should proceed incrementally, cautiously, and with humility, initially focusing only on those potential uses for which available knowledge has established an evidence-base and for which the balance of potential benefit and potential risk is carefully evaluated to ensure a high benefit-to-harm ratio.

To achieve this balance, the Commission concludes that any initial uses of clinical HHGE must meet *all four criteria* identified in this chapter. At present, it is only possible to define a responsible clinical translational path for applications of HHGE that fall into Category A or, possibly, a very small subset of Category B. For all other circumstances, additional considerations and lack of knowledge make it impossible today to properly evaluate the balance of risks and benefits, and the Commission is not currently able to describe a responsible translational pathway for clinical use.

Recommendation 3: It is not possible to define a responsible translational pathway applicable across all possible uses of heritable human genome editing (HHGE) because the uses, circumstances, and considerations differ widely, as do the advances in fundamental knowledge that would be needed before different types of uses could be considered feasible.

Clinical use of HHGE should proceed incrementally. At all times, there should be clear thresholds on permitted uses, based on whether a responsible translational pathway can be and has been clearly defined for evaluating the safety and efficacy of the use, and whether a country has decided to permit the use.

Recommendation 4: Initial uses of heritable human genome editing (HHGE), should a country decide to permit them, should be limited to circumstances that meet all of the following criteria:

1. the use of HHGE is limited to serious monogenic diseases; the Commission defines a serious monogenic disease as one that causes severe morbidity or premature death;

2. the use of HHGE is limited to changing a pathogenic genetic variant known to be responsible for the serious monogenic disease to a sequence that is common in the relevant population and that is known not to be disease-causing;

3. no embryos without the disease-causing genotype will be subjected to the process of genome editing and transfer, to ensure that no individuals resulting from edited embryos were exposed to risks of HHGE without any potential benefit; and

4. the use of HHGE is limited to situations in which prospective parents (i) have no option for having a genetically-related child that does not have the serious monogenic disease, because none of their embryos would be genetically unaffected in the absence of genome editing; or (ii) have extremely poor options, because the expected proportion of unaffected embryos would be unusually low, which the Commission defines as 25 percent or less, and have attempted at least one cycle of preimplantation genetic testing without success.

Chapter 4 sets out the elements that would be required for a responsible translational pathway toward initial uses of HHGE, in the event a country were to permit such uses.

4

A Translational Pathway to Limited and Controlled Clinical Applications of Heritable Human Genome Editing

This chapter identifies the elements of a responsible translational pathway for circumstances of heritable human genome editing (HHGE) that would fall into those described in Chapter 3 for potential initial uses: (1) prospective parents for whom all children would inherit the disease-causing genotype for a serious monogenic disease and who therefore have no alternative for having genetically-related offspring unaffected by the disease (Category A); and (2) prospective parents for whom some children would inherit the disease-causing genotype for a serious monogenic disease and who have poor likelihood of success through preimplantation genetic testing (PGT) (a very small subset of couples in Category B; see Chapter 3 for further details).

Chapter 4 specifies preclinical and clinical requirements that would need to be met to enable clinical evaluation of initial proposed uses of HHGE, should a country decide to permit such uses to be considered. A pathway toward clinical use of HHGE begins with a specific proposed use and includes three major stages:

1. development of a sufficient methodology and preclinical evidence of its safety and efficacy;
2. decision points and required approvals; and
3. clinical evaluation of a proposed use.

Each stage includes sub-components, as shown in Figure 4-1. This chapter describes these components and the requirements that would need to be met to proceed further. These pathway requirements pertain to genome

121

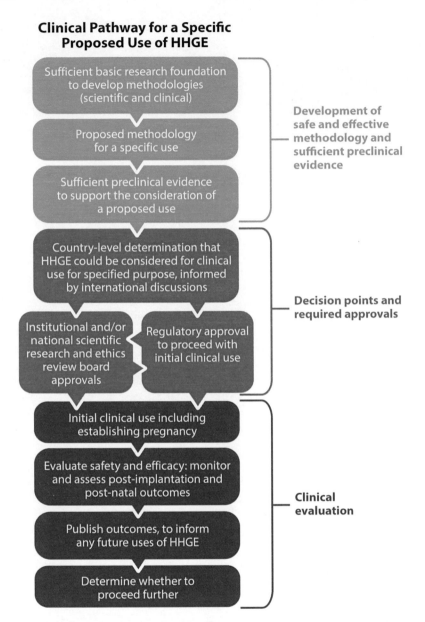

Clinical Pathway for a Specific Proposed Use of HHGE

Sufficient basic research foundation to develop methodologies (scientific and clinical)

Proposed methodology for a specific use

Sufficient preclinical evidence to support the consideration of a proposed use

Development of safe and effective methodology and sufficient preclinical evidence

Country-level determination that HHGE could be considered for clinical use for specified purpose, informed by international discussions

Institutional and/or national scientific research and ethics review board approvals

Regulatory approval to proceed with initial clinical use

Decision points and required approvals

Initial clinical use including establishing pregnancy

Evaluate safety and efficacy: monitor and assess post-implantation and post-natal outcomes

Publish outcomes, to inform any future uses of HHGE

Determine whether to proceed further

Clinical evaluation

FIGURE 4-1 A clinical translational pathway for a specific proposed use of HHGE. In this report, the Commission identified the elements that could form a translational pathway for cases of serious monogenic diseases in which all or a significant majority of the prospective parents' children would inherit the disease-causing genotype. The specific use of HHGE proposed for this pathway must therefore be one that falls within these categories of circumstances.

editing undertaken in human zygotes. Should in vitro–derived gametes ever be permitted as a reproductive technology, preclinical considerations for their use for HHGE are discussed later in the chapter.

As introduced in Chapter 1, important parallel processes of societal engagement also must occur throughout the pathway but are not the focus of this report.

The required elements of a responsible translational pathway are summarized in Box 4-1 and discussed in the chapter.

BOX 4-1
Essential Elements of a Responsible Translational Pathway Toward Initial Clinical Uses of HHGE

Basic Research Foundation: Undertake continued basic research to optimize genome editing technologies

Preclinical Evidence to Support a Proposed Use: Develop a proposed methodology for a specific use and obtain preclinical evidence
- Need for extensive research in cultured human cells and in zygotes of model organisms
  - Assessment of parental genomes
  - Testing of genome editing reagents in cultured parental cells
  - Testing of genome editing reagents in embryos of model organisms
- Preclinical testing in human embryos
  - Characterization of editing at the target site
  - Characterization of any off-target editing
  - Characterization of any mosaicism
  - Characterization of embryo development

Decision Points and Required Approvals: Obtain all required approvals, including those specified by national regulatory systems, and obtain informed parental consent

Undertake Clinical Evaluation of a Proposed Use
- Create genome-edited human embryos intended for transfer to establish a pregnancy
- Characterize human embryos intended for transfer

Evaluate Clinical Outcomes
- Monitor a resulting pregnancy
- Undertake longer-term monitoring and follow-up of any child born following HHGE
- Make information on decisions to permit the clinical evaluation of HHGE publicly available
- Evaluate information to inform future decisions about HHGE

CONTEXT FOR ANY HERITABLE HUMAN GENOME EDITING TRANSLATIONAL PATHWAY

As noted in Chapter 3, it is not possible to describe a generic translational pathway applicable to all uses of HHGE. Any translational pathway starts with the specific proposed use, which would involve making precise changes to a targeted sequence of DNA in the context of prospective parents wishing to have a genetically-related child without a particular disease. As emphasized in Chapter 3, the proposed clinical use also needs to be one that would fall within the set of circumstances for which the Commission was able to describe a translational pathway given the current state of scientific and clinical knowledge.

For any initial uses, HHGE would represent a new technological intervention in the assisted reproductive technology (ART) clinic, with only preclinical data with which to judge safety and efficacy. There will be information relevant to safety and efficacy that could only be obtained following evaluation in humans. As a result, the preclinical and clinical standards would need to be set very high for any initial human uses.

To meet this requirement for any initial human uses, the proposed use should be to change a pathogenic genetic variant known to be responsible for the serious monogenic disease to a sequence that is common in the relevant population and that is known not to be disease-causing. The disease would also need to be one that meets the Commission's definition of "serious" for the purpose of identifying an initial pathway toward HHGE. The Commission defines this as a life-shortening disease that causes severe morbidity or premature death.

BASIC RESEARCH FOUNDATION TO ESTABLISH SAFE AND EFFECTIVE GENOME EDITING METHODOLOGIES

As described in Chapter 2, current genome editing technologies are not sufficiently precise and specific to ensure safe and effective HHGE. Knowledge gaps remain in controlling and characterizing genome editing in human zygotes. Bringing the process of genome editing in zygotes to required levels of efficacy and safety will require substantial improvements in the editing and validation procedures themselves.

Basic Research as a Foundation to Develop Methodologies

Continued basic research is needed to expand understanding and control of genome editing in human zygotes. Continuing basic research on genome editing for purposes not linked to specific clinical uses will be very important for issues like design of the editing reagents for maximum efficiency and specificity; methods for detecting and quantifying the broad range of outcomes at both on- and off-target sites; enhancing desired editing outcomes—for example, by favoring homology-directed repair (HDR) over non-homologous end joining (NHEJ) if introducing a double-strand break is part of the methodology; and characterizing processes in human embryos that influence editing outcomes and may differ from those in somatic and cultured cells. Accumulating evidence will help to decide to what extent cultured cells, model organisms, or other surrogates can be used to confidently predict events in human zygotes.

Key elements that will be required to develop safe and effective methodologies for HHGE include the following.

Controlling On-Target Events

The inability to control events at the genomic target site constitutes a major limitation for HHGE. The majority of disease-causing mutations would require introducing the non-disease-causing sequence by copying from a provided template or from a non-disease-causing gene copy located on the homologous chromosome. Based on limited experience, this process of HDR is not efficient in human zygotes following a double-strand break in the DNA. In zygotes and in other cell types, the more common outcome of making a double-strand break is the introduction of sequence insertions and deletions (indels) via NHEJ. The NHEJ process could result in replacing one mutation with another, the consequences of which cannot be predicted or controlled. Such products would be deleterious in almost all instances; therefore, the ratio of DNA repair by HDR to NHEJ must be increased to achieve the desired outcome with high probability. As noted in Chapter 2, both base editing and prime editing largely avoid the risks associated with making and repairing a double-strand break; and both (particularly base editing) have shown promise in embryos.

The goal of HHGE would be to generate embryos that carry only a common, non-disease-causing sequence at both alleles of a gene. Creating one non-disease-causing allele would still be effective in the case of a recessive condition, while restoring both would also eliminate carrier status. A corollary of this goal is that alteration of a pre-existing non-disease-causing allele should be avoided. Zygotes that require genome editing must be identified prior to treatment. The latter may be possible through biopsy and testing of the first and second polar bodies (see Chapter 2) or future development of efficient genome editing methodologies for multi-cellular embryos that have already been genotyped.

Minimizing Off-Target Events

The ability to reduce the frequency of unintended genetic changes and to detect such changes when they occur has progressed significantly in recent years. For CRISPR-based genome editing, testing of various guide ribonucleic acids (gRNAs) for a particular target and making modifications to both the gRNA and the Cas protein have improved specificity. Similar advances have been made for the zinc-finger nuclease (ZFN) and transcription activator–like effector nuclease (TALEN) platforms. However, there are still challenges for detecting unintended sequence changes with high confidence in embryos. Analysis of off-target events arising from genome editing can be done by whole-genome DNA sequencing; however, current whole-genome sequencing (WGS) methods are not adequate for the accurate analysis of the small amount of genetic material that can be safely extracted from blastocyst-stage embryos intended for transfer to the uterus. In addition, WGS may not capture the full range of alterations that can occur. These could include large insertions and deletions or even whole or partial chromosome losses, which are difficult to detect with WGS or with standard polymerase chain reaction–based procedures.

Minimizing Mosaicism

Preventing mosaicism requires the ability to make the desired on-target modification with very high efficiency either in the one-cell zygote with restriction of editing activity to that stage or in all cells of embryos comprised of two or more cells. If genome editing continues beyond the first cell division, different cells in an embryo may carry different sequence changes at the intended target or at off-target sites. The effect of such mosaicism is difficult to predict, but it may pose serious risks by either failing to prevent disease due to target tissues having an insufficient number of appropriately edited cells or by introducing undesired mutations—particularly large copy number variants—at the target locus or elsewhere in a fraction of cells that could result in diseases related or unrelated to the targeted disease. Mosaicism poses particular challenges to verification. For an embryo destined for transfer, only a few trophectoderm cells can be safely removed from a blastocyst for molecular analysis. No current method can determine whether all cells of an embryo intended for uterine transfer carry exactly the same edits; it is even difficult to envision one that would. This means that preclinical research must establish procedures that only very rarely lead to mosaic embryos.

Evaluating the Physiological Effects of Genome Editing of Disease Alleles

Research on the short-term and long-term physiological and functional outcomes of editing disease alleles is needed to verify that a given intended edit is sufficient to prevent the disease phenotype and to provide reassurance that significant, unanticipated health effects would be unlikely to result from the genome editing process. Useful information may be obtained from human somatic editing of the same disease allele and potentially from the use of germline editing in other mammals—for example, to alter the animal-equivalent version of the human allele in cases where the human disease phenotype is accurately reproduced.

PRECLINICAL EVIDENCE TO SUPPORT A PROPOSED USE

Extensive investigation in a variety of experimental, preclinical contexts will be required prior to any attempt to establish a human pregnancy with an edited embryo.

Proposed Methodology for a Specific Use

Developing a proposed methodology that has been independently validated to be sufficient for the proposed use is an important part of the preclinical stage. The genome editing system (e.g., the combination of a Cas protein and gRNA designed to target a specific section of DNA) would need to address the issues above: controlling on-target editing, minimizing off-target events, and avoiding the generation of mosaic embryos. For any specific clinical use, the particular reagents and processes will need to be tested carefully at the particular genomic site and in the particular context as far as possible, as described below.

Sufficient Preclinical Evidence to Support Clinical Evaluation of the Proposed Use

To undertake HHGE through the use of zygotes, the genome-editing reagents would most likely be injected directly into oocytes concomitant with sperm or into zygotes immediately after fertilization. It is possible that, as has been observed for base editing (Zhang et al., 2019), treating embryos at the two-cell stage can also be effective. If HHGE were used in circumstances in which only some embryos were likely to carry the disease-causing genotype, the criteria for a translational pathway described in Chapter 3 would require a method to ensure that no embryos without the disease-causing genotype were subjected to the process of genome editing and transfer. Genotyping the polar bodies produced as an oocyte undergoes meiosis could identify the zygote's genotype in circumstances in which the maternal contribution is definitive (see Chapter 2). If polar body analysis would not be sufficient to determine the zygote's genotype, an alternate approach would be needed. One option could be editing an eight-cell or later embryo (the stage at which embryo biopsy and genotyping could be conducted). However, this would require the development of methodologies capable of effectively undertaking such editing. The goal for any of these circumstances is to produce embryos with a non-disease-causing genotype at the target sequence in all cells. Preclinical evidence for the proposed use of HHGE would need to be obtained from cultured human cells and from zygotes of model organisms before conducting preclinical experiments in human embryos.

What preclinical evidence can or should be collected depends on the genetic circumstances of the prospective parents, such as whether it is necessary to assess impacts on any non-disease-causing alleles that they might pass on. As described in Chapters 2 and 3, circumstances in which all embryos would inherit the genotype causing a serious monogenic disease include having one parent who is homozygous for an autosomal dominant disease-causing mutation or both parents homozygous for an autosomal recessive disease-causing mutation. In the latter case, no non-disease-causing allele would be present in cells of either parent or in their zygotes.

Need for Extensive Research in Cultured Human Cells and in Zygotes of Model Organisms

Each combination of a specific target gene and editing reagents would need to be evaluated, since each combination will present unique potential for on- and off-target events. To justify the design of the editing reagents proposed for potential clinical use, preclinical research in cultured human somatic cells from the prospective parents and in model organisms must include the following steps.

Assessment of Parental Genomes

Requirement: Obtain whole-genome sequences of the prospective parents using best practice protocols for investigating genetic disorders. Identify the exact sequence of the target mutation and surrounding genomic region. For a given combination of target and editing reagents, assess potential off-target sites based on these genomes.

Context: WGS is routinely used to identify new (*de novo*) mutations in offspring, as well as to establish the specific disease-causing genetic variant that parents with a history of genetic disease are at risk of passing on.

Examples of best practice protocols include the Deciphering Developmental Disorders study in the United Kingdom.[1]

Testing of Genome Editing Reagents in Cultured Parental Cells

Requirement: The following assessments need to be undertaken:
- For assessing on-target efficiency, test the editing reagents using cells from the parent(s) with the disease-causing mutation.
- For identifying sites at risk of off-target editing, test in cells from both parents.
- If a non-disease-causing allele is present in the genome of either parent, also test for any potential undesired editing of this allele.

Context: Testing in parental cells is important to allow for possible effects of genetic background on on-target and off-target outcomes. The information obtained from these assessments should be used to refine the editing reagents for efficacy at the intended target and to assess and minimize off-target mutagenesis. The cumulative frequency of off-target mutagenesis should not be significantly higher than the expected *de novo* mutation frequency. The cells could be primary cells from each parent, induced pluripotent stem cells (iPSCs), or embryonic stem cells (ES) derived from the parents by nuclear transfer. Because the adaptation to culture and the induction of pluripotency can both lead to the accumulation of novel mutations, testing in several independently derived lines is advisable.

Testing of Editing Reagents in Embryos of Model Organisms

Requirement: Test the efficiency of modifying the comparable target sequence in zygotes from a mammalian model organism. Use models incorporating humanized sequences—at least the sequence to be modified at the target and surrounding regions recognized by the editing reagent.

Context: Genome editing in mammalian zygotes differs from editing in somatic cells of the same species. Since some embryo-specific characteristics are likely shared among species, testing the editing reagents in zygotes of mammalian model organisms allows the characterization of the types of editing outcomes and the development of procedures to prevent and

---

[1] See https://www.ddduk.org/intro.html. The Deciphering Developmental Disorders study is funded by the Health Innovation Challenge Fund and the Wellcome Sanger Institute to analyze genomic information from "over 12,000 undiagnosed children and adults in the U.K. with developmental disorders and their parents" in order to better understand the basis of these disorders.

assess mosaicism. Relevant contextual factors in mammalian embryos include the methods of delivering the editing reagents, which are different in zygotes compared with cells in culture, and the DNA repair mechanisms active in zygotes compared with adult cells. While processes in human zygotes may differ from those in other mammalian zygotes, information obtained from such experiments can provide guidance for further testing in the human context.

Testing in mammalian zygotes is essential for refining the editing system prior to preclinical testing in human embryos. This testing is not designed primarily to evaluate phenotypic effects of editing the target sequence in the model organism. Animals such as mice can be used to generate sequences equivalent to the human disease-causing allele, so-called humanized alleles (Zhu et al., 2019). Evidence showing that the humanized disease-causing allele can be edited to the non-disease-causing allele would be essential to have prior to the earliest clinical uses of HHGE. If the humanized sequence cannot be produced in a mammalian model organism for some reason, then this disease allele should not be selected for an initial application of HHGE.

Preclinical Testing in Human Embryos

Preclinical testing in human zygotes must be undertaken to demonstrate that the genome-editing methodology proposed for clinical use provides high levels of efficiency, specificity, and safety. No other cell type can substitute for this stage of preclinical evidence. The preclinical testing of human embryos is conducted in a laboratory, and the embryos are never used to establish a pregnancy.

Only a limited number of human zygotes are available for experimental purposes, and the Commission recognizes that many jurisdictions do not permit the creation of human embryos for research. Nonetheless, thorough validation of the genome editing process prior to clinical use would require data from human embryos. To minimize generation of embryos specifically for experimentation, zygotes created through ARTs but not used by a couple to establish a pregnancy may be donated for use in laboratory research. While such zygotes would likely lack the specific disease-causing allele(s) being targeted for the proposed use of HHGE, testing them would generate information about potential off-target editing and could provide valuable guidance regarding zygote-specific processes involved in on-target editing. This option is limited by the fact that these stored zygotes will likely be at the G2 cell cycle stage—that is, later than would be subjected to genome editing—but may still be useful for approaches that address this stage or even ones in development for two-cell embryos. In addition, most in vitro fertilization (IVF) embryos are currently stored at even later stages and may not be useful at all.

For initial uses of HHGE, there will be certain types of information that could only be obtained following human clinical use. As a result, the standards for preclinical testing must be very high, and human zygotes for preclinical assessment of the editing methodology should be obtained that contain the disease-causing mutation. The efficacy of the editing reagents on disease-causing alleles carried by the prospective male parent can be tested in zygotes produced using his sperm to fertilize donated oocytes. In the case of disease-causing alleles transmitted by the female parent, care must be taken to avoid subjecting her to multiple rounds of hormonal stimulation and oocyte collection, with their attendant risks. If she is not at further risk from the IVF process and not of advanced age, a single round of stimulation and collection might be appropriate for the purposes of generating embryos for testing. A reasonable alternative would be to recruit a sperm donor who carries the same disease-causing mutation(s) and to use his sperm in conjunction with donated oocytes to produce zygotes for testing.

After a substantial knowledge base has been gained through experimentation on human embryos, it might be possible to identify alternative cell-types that reliably allow accurate prediction of the effects of genome editing in human zygotes. In such circumstances, it might become acceptable to use these cells as a surrogate for the preclinical tests involving human embryos required in this pathway. Rigorous scientific assessment of such models and their ability to substitute for evaluation in human embryos would be critical before using such alternative cell systems as the only source of preclinical evidence for HHGE. Over time, extensive preclinical testing of the ability of a particular editing methodology to correct a variety of targeted alleles might also become considered sufficient to conclude that it was not necessary to test the correction of each specific allele in preclinical human zygotes. Ongoing and independent scientific and technical reviews to assess knowledge gained and to consider whether it may be reasonable to make changes to preclinical standards required for the earliest human uses would be crucial (see Chapter 5 for a discussion of such oversight issues).

As noted above, preclinical testing can be complicated by the particular genetic circumstances of each couple. For example, when one or both prospective parents is a compound heterozygote for a serious genetic disease, more than one disease-causing allele will be present for that disease. This could pose a challenge to the design of gRNAs as they may only be able to edit one disease-causing allele and not others. Preclinical testing will need to examine the effects of the editing reagents on both the targeted allele and any other alleles present. In such circumstances, PGT on any edited embryo intended for uterine transfer will have to determine that at least one disease-causing allele has been changed to a non-disease-causing allele that is common in that population and that the other allele is unaffected.

Preclinical testing in human embryos must include the following steps.

Characterizing Editing at the Target Site

Requirement: The efficiency of the intended edit must be very high when measured in a cohort of treated human embryos. There must be no other sequence changes induced at the target, including insertions and deletions (indels).

Context: The goal of this testing would be to guarantee that the genome-editing methodology produced sufficient numbers of high clinical grade embryos with the desired edit before moving further toward to any clinical use. For a dominant disease, both zygotic alleles would need to exhibit a non-disease-causing sequence. For a recessive disease, although the restoration of a non-disease-causing sequence in one allele would prevent the disease, the editing frequency must be high enough that a high proportion of the available embryos are so modified. This testing could be done at any multicellular stage and must include testing for large deletions, chromosome loss, and other rearrangements.

Characterizing Any Off-Target Editing

Requirement: Compare parental genomes with whole-genome sequences obtained from the edited embryos or ES cells derived from these. Targeted sequencing should also be done for any particular off-target sites identified in preclinical research. There must be no detectable editing-induced off-target sequence changes. The incidence of *de novo* mutations, determined in conjunction with the sequence of the biological parents, must be in the range observed for unedited embryos, with no increase in the occurrence of single-nucleotide polymorphisms, indels, copy number variants, or chromosome rearrangements.

Context: Testing would be done at the blastocyst stage to provide sufficient DNA for this analysis and because this is the stage at which embryos would be transferred in the case of actual clinical use.

Identifying off-target sites from analysis of the cultured parental cells (see the first step under "Need for Extensive Research in Cultured Human Cells and in Zygotes of Model Organisms," above) would provide an initial indication of high-risk off-target sites in the embryo, allowing attention at this stage to focus on regions where off-target editing has previously been observed or might be anticipated to occur.

Characterizing Any Mosaicism

Requirement: All cells of the embryo must have the same on-target sequence (i.e., no mosaicism) as shown by analysis of multiple individual cells. The sequences of the intended target and any high-risk off-target sites should be determined for each individual cell, or as many as feasible, in the early-stage embryo.

Context: If not all cells have been successfully edited, it is possible that the target organ(s) in the offspring, and/or later in the adult, will not be completely disease-free.

Characterizing Embryo Development

Requirement: The genome-edited embryos must proceed through normal development *in vitro* to the blastocyst stage, meeting milestones with comparable efficiency to unedited embryos. Cellular and molecular features of genome-edited embryos should be comparable to unedited embryo controls and have aneuploidy rates no higher than expected based on standard ART procedures.

Context: The goal of such testing is to ensure that the genome editing does not negatively affect normal embryo development. Developmental characterization in genome-edited embryos would be compared to expected embryo development based on what is known from the use of unedited embryos in IVF. This assessment could be continued up to the 14-day limit currently permitted for human embryo culture in many countries.

Examples of best practice protocols include those used by the Newcastle Fertility Centre at the International Life Science Centre and others in the preclinical evaluation of human embryos that had undergone mitochondrial replacement techniques (MRT). In collaboration with the Crick Institute, investigators at Newcastle analyzed the cell lineages that were present in blastocyst-stage human embryos that had undergone MRT, to ensure that all expected cell lineages were present. They also performed single-cell transcriptome analysis to check for the expected patterns of gene expression (Hyslop et al., 2016).

Only if all of these preclinical requirements are met and validated by independent expert opinion should the use of edited embryos in a clinical setting be contemplated.

Additional Consideration for Genome Editing in Which Not All Embryos Would Inherit the Disease-causing Mutation

Requirement: Development of a genome editing methodology capable of safely and efficiently editing an eight-cell or later embryo may be required.

Context: If all of the embryos that can be produced by the prospective parents will inherit the disease-causing mutation, genome editing could be undertaken around the time of fertilization. If only some of the embryos will carry the disease-causing mutation, identifying which are affected would be necessary prior to editing. For genetic circumstances in which only the maternal genetic contribution needs to be known, identifying the oocyte genotype through polar body analysis may be sufficient. In other circumstances, in which both maternal and paternal genetic contributions need to be known, an embryo biopsy would be required. For any such envisioned uses, the preclinical development and testing of a genome editing methodology capable of safely and effectively editing embryos post-genotyping would be required.

DECISION POINTS AND REQUIRED APPROVALS

Several important approvals must be received before any clinical use of HHGE could be undertaken.

Determination That Heritable Human Genome Editing Could Be Considered for Clinical Use in a Country

As described in Chapter 1, a country must first allow the consideration of HHGE for the proposed clinical use. This decision making will not only include information on preclinical evidence of an appropriate genome-editing methodology, but also include societal engagement and input. The clinical use of HHGE remains illegal in many countries; many others have not yet established oversight systems by which HHGE would be regulated, should it be permitted. It would be important for any clinical uses of HHGE to take place only in the context of a regulated environment (see Chapter 5 for additional discussion).

Appropriate Review Board and Regulatory Approvals

A proposal to clinically evaluate a particular use of HHGE would require submission of information on the proposed disease and genomic target, preclinical evidence, and clinical protocols to relevant institutional and national advisory bodies for science and ethics. Appropriate approvals would need to be obtained as a result of these reviews. A proposal with the required supporting preclinical evidence and protocols must also be reviewed and permitted by the appropriate national regulatory authorities. Only if such approvals are obtained could the initiation of a pregnancy with edited embryos for the proposed use be undertaken.

Informed Consent from Prospective Parents

For any clinical evaluation of HHGE, the prospective parents must be informed about the procedures and projected outcomes as thoroughly as possible, and they must give their consent to the intervention. As a requirement of informed consent is that prospective parents are given detailed information about the nature and risks of HHGE, it is premature to establish specific protocols at this time. Instead, general guidelines are presented of the principles and procedures that should be considered in the informed consent process. Due to the technical nature of genome editing, this will require extensive discussions in most cases. Prospective parents would require clinical assessment and counseling, by people with no conflict of interest regarding the outcome. Counseling would need to include the presentation of all reproductive options, including the risks, benefits, and degree of unknowns associated with each, with opportunities for prospective parents to consider the implications of the choices available to them. Reproductive advice would need to cover all aspects of ARTs, including a discussion of IVF, PGT, and any interventions used for prenatal evaluation. The prospective parents would also be asked to give consent to fetal monitoring and to reasonable post-natal monitoring and assessment. Assessment and counseling would need to consider mental health as well as physical health, both in parents and in resulting offspring, and the prospective parents' ability to care for children born. Psychological support would also need to be available throughout the consent process.

In addition to meeting standard criteria for informed consent, because HHGE would represent a novel technology without a history of clinical use, care would need to be taken for any initial human uses not to engender or to be influenced by excessive optimism. It will be essential for those leading consent discussions to have no conflict of interest regarding the outcome of HHGE and fully understand the mechanisms, procedures, and risks involved.

CLINICAL EVALUATION OF THE PROPOSED USE

Once all of the required preclinical evidence had been assembled, indicating that a suitable methodology was available, and all of the appropriate regulatory reviews and approvals had been completed, a genome-edited human embryo might be generated with the aim of establishing a pregnancy.

The required clinical elements include the following.

*Identify the Genotype (for Circumstances in Which
Not All Embryos Are Expected to Carry the Disease-Causing Genotype)*

Requirement: Identify the genotype of human oocytes and/or embryos prior to genome editing.

Context: This element may be required to ensure that only genetically affected embryos undergo genome editing. Sufficient genotype identification could be obtained through polar body biopsy in certain circumstances; in others, an eight-cell or later embryo biopsy may be required. In such cases, a genome editing methodology capable of editing a multicellular embryo must be established during the preclinical phase.

*Create Genome-Edited Human Embryos Intended
for Transfer to Establish a Pregnancy*

Requirement: Best practice standards for the relevant genome-editing and ARTs would need to be followed in obtaining the parental gametes and creating a genome-edited zygote.

Context: The medical center performing the creation of the zygote, introduction of the genome-editing reagents, assessment of the clinical suitability of the resulting embryos, and eventual transfer to establish a pregnancy would need to have the appropriate qualifications, experience, and demonstrated competences according to the regulatory requirements of its country and would need to adhere to professional best practice guidelines. Best practice standards for consistency and quality control would also need to be followed for all reagents and procedures.

Characterize Human Embryos Intended for Transfer

Requirement: Perform an embryo biopsy to collect cells from the trophectoderm of blastocyst-stage embryos and perform PGT to confirm the presence of the precise on-target edits, the absence of detectable off-target mutations, and no evidence of mosaicism.

Context: As detailed above, extensive preclinical evidence must demonstrate that a methodology is consistently able to deliver human embryos in which every cell has the appropriate genetic features following genome editing. As a result, a trophectoderm biopsy of an embryo intended for transfer would be expected to reliably correlate with the rest of the embryo.

Evaluation of Outcomes Including Safety and Efficacy

Should a pregnancy be established with a genome-edited human embryo, it will be important to evaluate any negative effects during the prenatal period, as well as to assess physical and psychological outcomes of any child born following HHGE. It will also be important that information on the clinical outcomes of HHGE, including any detected negative effects, be collected and assessed to inform the understanding of safety and efficacy.

Monitor a Resulting Pregnancy

Requirement: Careful monitoring of a resulting pregnancy with a genome-edited embryo is strongly recommended.

Context: Following transfer of a genome-edited embryo to establish a pregnancy, prenatal monitoring is crucial to detect any fetal abnormalities or other issues arising during the pregnancy. Should prenatal testing identify genetic or physical anomalies, counseling for the parents will be important. Such prenatal monitoring would be strongly recommended but is the choice of the mother.

Undertake Longer-term Monitoring and Follow-up

Requirement: Longer-term monitoring and follow-up of a child born following HHGE is essential, and should include:
- obtaining consent from the parents, and later from the child, for monitoring immediately after birth and at specified intervals thereafter extending into adulthood, which must be done by competent professionals and include both physical and psychological aspects; and
- using assessment tools that have been validated and standardized in an international context and, if appropriate, that have versions available across the lifespan.

Context: It is important for the health of the individual born as a result of HHGE, as well as any children that they have, that such individuals continue to be assessed for adverse genetic or health outcomes. If adverse outcomes are identified, the individual concerned should be informed of them if they so choose and offered genetic counseling.

Make Information on Decisions to Permit the Clinical Evaluation
of Heritable Human Genome Editing Publicly Available

Requirement: Each country would need to make public the details of any approved applications to clinically evaluate HHGE. Information that would need to be made available includes the genetic condition for which HHGE had been allowed, the associated laboratory procedures that would be used, and the national bodies that would be providing oversight.

Context: Making such information publicly accessible would be important for ensuring transparency about any potential uses of HHGE being contemplated, the evidence base on which decisions had been made, and oversight responsibilities. However, information made available would need to protect family identity.

Evaluate Information to Inform Future Decisions
about Heritable Human Genome Editing

Requirement: It would be vital to publish in peer-reviewed scientific journals the procedures involved and outcomes of any clinical evaluation of HHGE.

Context: Such information would contribute to ongoing national and international discussions on the safety and efficacy of HHGE. In conjunction with further extensive societal engagement, such information would also contribute to any decisions about whether to consider the clinical evaluation of HHGE for other uses in Categories A and B, according to the translational pathway identified in this report, whether or how to modify any of the preclinical or clinical requirements laid out in this translational pathway, or potentially whether to consider evaluating uses that would fall into other categories of potential uses described in Chapter 3.

HERITABLE HUMAN GENOME EDITING USING IN VITRO STEM CELL–DERIVED GAMETES: WHAT A POTENTIAL TRANSLATIONAL PATHWAY WOULD ENTAIL

Chapter 2 describes the prospect of genome editing in human gamete precursors by two approaches: editing gamete precursor cells, such as spermatogonial stem cells (SSCs); and editing pluripotent stem cells followed by differentiation into functional gametes *in vitro* (in vitro–derived gametogenesis). At present, procedures for generating functional human sperm or oocytes by these methods are not available, so this technology is not available for clinical use. It should be emphasized that methodologies for using SSCs, (iPSCs), or nuclear transfer embryonic stem cells (ntESCs), even if safe and efficient, would need to be permitted for use in assisted reproduction in the absence of genome editing, independently of considering their use in combination with genome editing. Such approval would need to involve extensive public consultation given its societal implications. Because this technology is not yet available for approval in any clinical setting, it would be premature to describe a translational pathway that uses it to create heritable genomic changes. Nevertheless, the section below describes preclinical and clinical considerations that would be relevant to such a pathway were it ever to be feasible.

Clinical availability of in vitro–derived gametes for assisted reproduction, especially female gametes that are usually only available in small numbers, would eliminate the need for HHGE in all monogenic disease circumstances except Category A. This is because it would be possible to generate and genetically test sufficiently large numbers of embryos such that identification of those that do not have the disease-causing genotype would be practically assured. Moreover, since the availability of such unaffected embryos would no longer be limiting, it would be possible to select those of the highest clinical grade for transfer to the prospective mother.

This strategy would not be available in the context of Category A, because no in vitro stem cell–derived gametes, and therefore no embryos, could be produced that lack the disease-causing genotype. In this case, undertaking HHGE would require genome editing of patient-derived stem cells *in vitro*, resulting in the production of cells lacking the disease-causing genotype. Generation of functional gametes from such edited stem cells would have several advantages over genome editing in zygotes. A significant advantage arises from the fact that editing would be done in cultured cells, where methods for making and evaluating specific modifications and for analyzing genomic and epigenetic profiles are well documented, although there is no consensus on which methods are best. The per-cell editing efficiency would not have to be particularly high, because treated cells can be thoroughly characterized for on- and off-target editing to find those with only the desired changes, expanded into a pool of correctly edited cells, and then differentiated into functional gametes. Finally, mosaicism would not be an issue when a single sperm derived from an edited SSC, iPSC, or ntESC is used to fertilize an egg.

However, the use of in vitro stem cell–derived gametes could have disadvantages. The precursors of such gametes would have gone through extensive adaptation to and expansion in cell culture. During this time, *de novo* mutations could accumulate at levels comparable to the spontaneous germline mutation rate *in vivo* (Wu et al., 2015). Some cells might be selected for properties, including both genetic and epigenetic differences, that promote their ability to replicate in culture conditions, which could have unknown effects if used clinically. Each batch of cells may acquire a unique set of mutations, unlike the more specific off-target mutations potentially induced by genome editing. It is not clear how to evaluate such genetic and epigenetic changes, were they to be unavoidable, for possible effects on embryonic, fetal, and post-natal development.

Were approaches that rely on the use of in vitro stem cell–derived gametes to be permitted for clinical use as a reproductive technology, their use for HHGE would still be subject to the same ultimate tests of safety and efficacy outlined above for genome editing in zygotes. However, there would also be some special considerations. For example, there would be no need to test the resulting embryos for mosaicism.

Before any clinical use to create a human embryo for transfer to the uterus, preclinical research for genome editing approaches using in vitro stem cell–derived gametes would include the following.

- *Extensive research in human cells to develop and optimize the genome editing reagents.* For editing in gamete precursor cells, such as SSCs, comparative genetic analysis with uncultured SSCs from the prospective father would be required. This would allow the development of effective editing reagents for on-target efficiency of the desired edit and absence of other on- and off-target changes. For editing in iPSCs/ntESCs that would subsequently be differentiated into functional male or female gametes, similar comparative analyses would be undertaken with unedited parental iPSCs/ntESCs to optimize the editing reagents.
- *Isolation of individual cell lines and testing to characterize on- and off-target events and epigenetic profiles.* Cell lines derived from iPSCs/ntESCs that had undergone genome editing would need to be examined thoroughly by WGS for mutations acquired and selected for during establishment and growth in cell culture. Selected lines would need to have only the desired edit at the intended target and no undesired modifications elsewhere in the genome as a result of the editing. Epigenetic and gene expression profiles should also be examined to better understand whether the editing reagents had affected these.
- *Differentiation of the correctly edited iPSCs/ntESCs into functional gametes.* As discussed in Chapter 2, protocols have been developed that permit the derivation of functional gametes from mouse pluripotent stem cells *in vitro*. Similar protocols are being developed in humans, but there are major challenges that still need to be overcome, not least that of ensuring normal meiosis in such cells. Continued research in this area will be of vital importance, not only for the development of in vitro–derived gametes that can be used

safely, but also for the insight such research will allow into human gametogenesis and abnormalities thereof associated with infertility.

- *Characterization of gametes before their use in IVF.* This would require comparison with parental gametes not derived from cultured cells in order to assess genetic and epigenetic properties of the edited gametes, including additional WGS to examine potential genome changes acquired during the gamete differentiation process. Variation between the transcriptomic and epigenomic properties of individual gametes could be assessed through single-cell approaches (Hermann et al., 2018).

- *Testing the functionality of genome-edited gametes.* The ultimate test of any gamete generated *in vitro* will be its ability to generate an embryo with general features that are essentially indistinguishable from an embryo generated using conventional gametes. For example, it will be necessary to demonstrate that male gametes derived from either unedited or genome-edited precursor cells *in vitro* are able to effectively fertilize an oocyte and that resulting embryos reach normal developmental milestones to the blastocyst stage. The genome or epigenome of a single sperm that contributed to any one embryo may not be representative of the genomes characterized in bulk produced by this method of genome editing, so characterization of genomes and epigenomes from individual embryos generated from in vitro–derived gametes will be very important. ES cell lines could be derived from such embryos to confirm by high-quality genome sequencing that such cells did not differ from parental genomes at off-target sites. Finally, conducting genetic testing on cells obtained from a blastocyst biopsy would be important in the clinical phase, prior to transfer of any embryo, at least for the intended genomic target.

CONCLUSION AND RECOMMENDATIONS

Any responsible pathway to clinical application of HHGE would need to include clear and strict criteria in technical capabilities, in the acceptable evaluation of safety and efficacy, and in oversight standards. The Commission's recommendations set out the components that would be required for a responsible translational pathway and are provided below.

Scientific Validation and Standards for Any Proposed Use of Heritable Human Genome Editing

Evidence from preclinical research would be required to establish that HHGE may be safe enough to consider evaluating in first-in-human clinical applications. Once the preclinical requirements set out in the translational pathway have been met, it may become appropriate to proceed with clinical interventions, subject to required approvals, informed consent, and ongoing review and monitoring. Each specific clinical use would need to be carefully considered in its own right. Even with preclinical evidence, there will still be unknowns concerning safety and efficacy that could only be revealed and resolved by the long-term monitoring of individuals born following the use of HHGE.

> **Recommendation 5**: Before any attempt to establish a pregnancy with an embryo that has undergone genome editing, preclinical evidence must demonstrate that heritable human genome editing (HHGE) can be performed with sufficiently high efficiency and precision to be clinically useful. For any initial uses of HHGE, preclinical evidence of safety and efficacy should be based on the study of a significant cohort of edited human embryos and should demonstrate that the process has the ability to generate and select, with high accuracy, suitable numbers of embryos that:
> - have the intended edit(s) and no other modification at the target(s);
> - lack additional variants introduced by the editing process at off-target sites—that is, the total number of new genomic variants should not differ significantly from that found in comparable unedited embryos;
> - lack evidence of mosaicism introduced by the editing process;
> - are of suitable clinical grade to establish a pregnancy; and
> - have aneuploidy rates no higher than expected based on standard assisted reproductive technology procedures.

> **Recommendation 6**: Any proposal for initial clinical use of heritable human genome editing should meet the criteria for preclinical evidence set forth in Recommendation 5. A proposal for clinical use should also include plans to evaluate human embryos prior to transfer using:
> - developmental milestones until the blastocyst stage comparable with standard in vitro fertilization practices; and
> - a biopsy at the blastocyst stage that demonstrates
>   o the existence of the intended edit in all biopsied cells and no evidence of unintended edits at the target locus; and
>   o no evidence of additional variants introduced by the editing process at off-target sites.

If, after rigorous evaluation, a regulatory approval for embryo transfer is granted, monitoring during a resulting pregnancy and long-term follow-up of resulting children and adults is vital.

Future Developments Affecting Reproductive Options

Genome editing of gamete precursor cells or editing of pluripotent stem cells followed by *in vitro* differentiation into functional gametes could represent alternative methods of undertaking HHGE. However, the technologies to develop human gametes from cultured cells are still under development and are currently unavailable for clinical use.

Recommendation 7: Research should continue into the development of methods to produce functional human gametes from cultured stem cells. The ability to generate large numbers of such stem cell–derived gametes would provide a further option for prospective parents to avoid the inheritance of disease through the efficient production, testing, and selection of embryos without the disease-causing genotype. However, the use of such in vitro–derived gametes in reproductive medicine raises distinct medical, ethical, and societal issues that must be carefully evaluated, and such gametes without genome editing would need to be approved for use in assisted reproductive technology before they could be considered for clinical use of heritable human genome editing.

National and International Governance of Heritable Human Genome Editing

A responsible translational pathway toward potential clinical uses of heritable human genome editing (HHGE) requires that national and international governance foundations be in place prior to any clinical use. Chapter 5 discusses the elements that would need to be part of such systems. This chapter begins by discussing how HHGE intersects with, and poses challenges for, current oversight systems for medical technologies. The chapter then describes the mechanisms that a country would need to establish to ensure responsible oversight of any future clinical uses of HHGE. Finally, the chapter emphasizes the need for international coordination around developments that affect HHGE. The chapter does not delve into detail on how national and international governance systems for HHGE would ultimately be implemented by countries and by the international community—ongoing dialogues including the work of the World Health Organization's (WHO's) Expert Advisory Committee are exploring this area in greater depth. However, this chapter concludes with recommendations for core components of these efforts.

A RESPONSIBLE GOVERNANCE SYSTEM FOR HERITABLE HUMAN GENOME EDITING

HHGE would entail a form of assisted reproductive technology (ART) used to generate an embryo with an altered genome with a view to establishing a pregnancy. A governance system for the use of HHGE would need to include the ability to oversee all stages of the translational pathway described in Chapter 4. These stages include basic and preclinical research to develop methodologies for HHGE that can sufficiently control and characterize the effects of genome editing; national legislative, advisory, and regulatory decision making charged with determining whether a clinical use of HHGE could be considered; and evaluation of outcomes resulting from any clinical use of a genome-edited embryo to establish a pregnancy.

Considerations for Societal and Stakeholder Engagement on Heritable Human Genome Editing

Prior to the clinical use of HHGE in any country, one important requirement is for public engagement on whether it would be acceptable to use HHGE in that country and, if so, for what purposes and with what governance mechanisms. Genome editing in human embryos should not proceed past preclinical laboratory research unless it is deemed acceptable by a country and unless there are approvals by the relevant bodies to consider it for potential clinical use. The question of precisely how such discussions should proceed was beyond this Commission's charge; however, presentations and submissions to the Commission's call for evidence emphasized a number of additional points to inform future deliberations (see Box 5-1).

HERITABLE EDITING IN THE CONTEXT OF CURRENT REGULATORY SYSTEMS

A governance system for HHGE would share similarities with the oversight structures that currently guide appropriate conduct in other areas of biomedical research and clinical practice. Because HHGE entails the use of genome editing technologies as a form of assisted reproduction to enable prospective parents to have a child with an altered genome, it shares some characteristics with existing oversight systems in both somatic gene therapies and ARTs. However, the clinical use of HHGE would also pose challenges to current systems.

How Heritable Human Genome Editing Would Relate to the Regulation of Gene Therapies

Many somatic cell gene therapies currently undergoing clinical development rely on using genome editing technologies. Somatic cell gene therapies have a history of highly regulated oversight in the countries in which they have been carried out, including the United States, Japan, China, India, and countries in Europe. In the United States, the European Union, and China,

BOX 5-1
Societal Considerations to Inform Future
Discussions about HHGE[a]

Engagement with Genetic Disease, Disability, and Minority Communities

- **It is critical to engage directly with people who have conditions that might be considered for HHGE.** Views on HHGE among genetic disease and disability community members differ. Attitudes reported in a 2016 consultation by Genetic Alliance U.K. ranged from welcoming the potential ability to eradicate a condition, to serious reservations, to the view that a genetic condition is a fundamental part of the person's identity (Genetic Alliance UK, 2016). Communities expressing concerns include the deaf community and the autism community. Many submissions to the Commission's call for evidence emphasized that any demand for HHGE must come from communities of people who are living with the particular condition under consideration.

- **It is critical to recognize historical experiences with stigmatization and eugenic practices concerning disease and disability.** Concerns expressed by respondents over any use of HHGE included that it will "undermin[e] the equality of value and worth of all human persons in society." Other concerns expressed included that the development of HHGE could reduce the accommodations a country provides to people having genetic conditions.

- **It is critical to engage with other communities whose voices have not always been considered in medical decisions, including minority and indigenous communities.** For example, the incidence of sickle cell disease (SCD) in the African American community is substantially higher than in the U.S. population overall. It would be technically possible to consider HHGE to prevent inheritance of SCD, but past unethical and abusive medical conduct has left a legacy of mistrust of the medical establishment for this community. It would be essential to extensively and systematically engage with and incorporate input from African Americans before advancing toward any clinical use of HHGE for SCD.

Engagement with Civil Society

- **A need for discourse among civil society about human genome editing.** As stated by one respondent, "society must have the opportunity to shape the way in which the science develops." There are diverse levels of understanding about the meaning of scientific terms such as "genome," "somatic cell," and "germline"; what types of genome editing uses are currently being developed; and how HHGE would be undertaken should it be permitted. As a result, there is a role for education to underpin informed public consultations.

continued

BOX 5-1 Continued

- **A need to include an expansive array of topics in societal discussions of HHGE.** The focus of public engagement and civil society discussions will need to be on more than scientific and clinical dimensions and will need to include diverse voices as well as expert input from the humanities, social sciences, ethics, and faith communities. Issues that will need to be debated by a country include the potential for HHGE to prevent disease transmission; the implications of HHGE for exacerbating inequities and social justice concerns; the value placed on genetic relatedness of a child or on parental freedom to pursue reproductive preferences; potential social and psychological consequences for parents, children, and the wider family; privacy considerations; and others. Some respondents stated that HHGE had no possible path to legitimate use, while others could envision its use in certain circumstances, and still others noted that it currently conflicts with existing laws and international treaties. All of these issues will need to be openly debated by a given country.

- **A need to develop and support processes by which societal discourse can be undertaken.** This would include how to undertake societal engagement, how to engage diverse views, and how to support and sustain such efforts at national and international levels. It would be valuable to draw on expertise from the social sciences to develop effective engagement strategies.

- **A need for transparency and accountability associated with the development and potential use of HHGE.** Transparency can give legitimacy to decisions about HHGE and would include making available information on what evidence exists on the safety and efficacy of HHGE technologies, how (and by whom) this evidence is assessed and how (and by whom) decisions are made about whether HHGE could be undertaken, and the outcomes of any clinical use of HHGE. This information needs to be regularly updated.

---

[a] The information in this box is based on submissions to the Commission's call for input and presentations and comments during public information-gathering sessions.

for example, somatic genome editing is regulated primarily using the frameworks established for prior generations of gene therapies (NASEM, 2017).

A number of clinical trials based on somatic genome editing have been initiated in the United States. The regulatory process involves the institutional reviews required for human clinical trials as well as additional institutional oversight by biosafety committees and federal review. Federal oversight includes requirements for prospective approval from the national regulatory authority, the U.S. Food and Drug Administration (FDA), via an Investigational New Drug license or its equivalent. Once clinical trials commence, centralized reporting of adverse events and longitudinal data collected during the clinical trial phases are required for submission of an application to the FDA for approval as a therapeutic to be marketed in a clinical context.

Other countries have similar regulatory systems intended to ensure the safety and efficacy of somatic gene therapies tested or approved for use in humans. Regional organizations, such as the European Medicines Agency, promote the development of scientific guidelines in areas such as gene and cell therapy products, and there are other ongoing international dialogues aimed at improving the consistency of somatic gene therapy regulations.[1] In addition, countries including Chile, Colombia, Mexico, and Panama have incorporated explicit prohibitions on the use of somatic genome editing for purposes that might be perceived as enhancement (Abou-El-Enein et al., 2017; NASEM, 2017). While somatic genome editing shares a similarity with heritable genome editing in that both types of uses rely on genome-editing technologies, there are important differences that challenge the applicability of its regulatory frameworks to HHGE.

Somatic therapies fit within an oversight paradigm in which medical interventions are developed and deployed to treat an existing patient with a genetic condition. The effects of the editing are limited to that individual's cells and tissues and are not inheritable, and the largely individual-level harms and benefits can be assessed and explained as part of gaining informed consent. HHGE, on the other hand, provides a reproductive option for prospective parents to have a potential future child without transmitting a disease-causing genotype. The heritable genomic alteration and the potential harms, benefits, and uncertainties that arise may affect not only that child but also any offspring of that child, raising societal concerns and leading to effects that may not be apparent until subsequent generations.

How Heritable Human Genome Editing Would Relate to the Existing Regulation of Assisted Reproductive Technologies

As noted above, HHGE would constitute a form of ART, and ARTs have a very different history of regulatory oversight from that of somatic gene therapies. Laws regarding the use of ART vary substantially among countries. While there are important lessons to be gleaned from such regulatory experiences (Cohen et al., 2020), this variation will make it difficult to achieve coordinated oversight of HHGE using current ART regulatory systems.

---

[1] See http://www.iprp.global/working-group/gene-therapy.

A survey conducted in 2018 by the International Federation of Fertility Societies (IFFS), that spanned 89 of the 132 countries believed to be offering ART, found that 64 percent of countries that responded to the survey had legislation regulating the use of these technologies, with a focus on licensing clinics, physicians, and laboratories. Penalties for violating regulations ranged from admonishment to imprisonment, with the most frequently used sanctions reportedly being financial penalties, loss of license, and threat of criminal prosecution (IFFS, 2019).

One of the ARTs most relevant to the discussion of HHGE is pre-implantation genetic testing (PGT), in which cells removed from an early embryo created through in vitro fertilization (IVF) are genetically analyzed and only embryos having specified genotypes are transferred to establish a pregnancy. The majority of countries responding to the IFFS survey reported that they allow PGT for prevention of monogenic disease. These countries were split almost equally with respect to whether laws or regulations permitting the use of PGT were accompanied by further guidelines restricting how it could be used.

Research analyzing national approaches to the use of PGT found that they were typically based on the seriousness of a condition (Isasi et al., 2016). For example, Mexican legislation prohibits PGT for any purpose other than "the elimination or reduction of serious diseases or defects," while other countries require a "substantial risk" of the disease occurring or that the disease is "untreatable" or "incurable" (Isasi et al., 2016). The United Kingdom is an example of a country that utilizes "seriousness" in its evaluation of PGT applications and where the use of IVF with PGT is permitted only to prevent specific genetic conditions that have been approved by the Human Fertilisation and Embryology Authority. Its list of permitted conditions now totals more than 600 and includes the use of PGT to select an embryo that is immunologically matched to a sibling with a disease (savior sibling PGT).[2] In France, only couples at high risk of having a child affected by a "particularly serious and not curable" genetic disease are allowed to use PGT under the national public health code, with oversight by the country's biomedical agency (Agence de la Biomédecine). Requests for PGT are evaluated on a case-by-case basis rather than against a list of allowed uses. A review committee at each major reproductive medicine center evaluates the requested use and reports annually to the Agence de la Biomédecine on its decisions. This enables retrospective analysis of the criteria by which such requests are judged, which take account of factors such as risk of disease, anticipated disease manifestation, and family medical history. In China, regulations on ARTs were published by the

---

[2] See https://www.hfea.gov.uk/treatments/embryo-testing-and-treatments-for-disease/approved-pgd-and-ptt-conditions/.

Ministry of Health in 2001 and were amended in 2003 into a document titled "Technical Standards, Basic Requirements, and Ethical Principles on Human ARTs and Related Technologies and Human Sperm Bank." Any medical institutions permitted to carry out human ARTs are required to meet these regulations and standards and to obtain an approval certificate from the Ministry of Health. Medical institutions offering these technologies are required by law to have ethics committees, which review certain proposed methods or some specific cases. PGT has been used for those couples that are at high risk of having a child with single-gene disease, chromosome disorders, or sex-linked genetic disease, but it is not allowed for sex selection. In the United States, ARTs are offered in the context of the practice of clinical medicine without a requirement for regulatory approval. There are no federal restrictions on the conditions for which PGT can be used. Instead, PGT use is guided by any state laws that may restrict uses, professional guidelines, and the choices of clinicians and prospective parents (Bayefsky, 2016, 2018).

Mitochondrial replacement techniques (MRT) represent a novel form of ART, which are currently permitted for clinical use in the United Kingdom. The approach taken by the United Kingdom to develop a translational pathway and oversight regime for this technology can provide an informative model to help guide the development of national oversight systems relevant to HHGE. As described in Chapter 1, the characteristics include a controlled step-wise process under the auspices of appropriate national regulators; limitation to cases involving parents wishing to have a genetically-related child unaffected by serious disease; limited licensure to use in single cases rather than blanket approval, with ongoing review before subsequent licenses are issued; a comprehensive informed consent process; long-term follow-up of offspring; and prohibition of uses beyond the permitted indication.

Lessons Applicable to the Creation of an Oversight System for Heritable Human Genome Editing

As with other medical technologies, an oversight system for HHGE would need to address all stages of a research and clinical translation pathway. Because multiple actors contribute to any translational pathway, responsibilities at individual, institutional, national, and international levels will be required. Investigators and clinicians will need to adhere to relevant norms, guidelines, standards, and policies. For example, these may include or draw on policies developed for governance of gene therapies and for governance of ARTs. Well-specified processes will need to be established for institutional boards to review clinical use protocols, including appropriate protections for participants. Prior to the initiation of any clinical use, approvals will be required from relevant national advisory bodies and national regulatory authorities that assess the context of proposed use, preclinical evidence, clinical protocols, and plans for follow-up. Processes will need to be implemented for national and international discussion, coordination, and sharing of results on relevant scientific, ethical, and societal developments impacting the assessment of HHGE's safety, efficacy, and societal acceptability (see Box 5-2).

Legal and Regulatory Frameworks

The legal and regulatory status of HHGE varies considerably among countries. HHGE is currently prohibited by law in dozens of countries

---

BOX 5-2
Experience Conducting Independent Assessments of Safety and Efficacy of Mitochondrial Replacement Techniques

The pathway toward clinical use of MRT in the United Kingdom included detailed assessments in 2011, 2013, 2014, and 2016 of the state of the science and preclinical evidence on safety and efficacy (HFEA, 2011, 2013, 2014, 2016). This experience, described below, can inform the creation of systems for regular reviews of developments relevant to HHGE. As can be seen from the chronology, multiple reviews were undertaken as knowledge progressed, with evidence requested by one panel being generated by the scientific community and reviewed by a subsequent panel. The latest evidence on preclinical safety and efficacy also informed recommendations on which patients might be considered for initial human uses and what types of clinical follow-up and outcome assessments should be undertaken.

The MRT Scientific Assessments
Over the course of four scientific reviews, expert panels examined preclinical data on the use of MRT in model organisms and in human research embryos, scrutinizing both published and unpublished data. One of the key issues evaluated was the ability to produce animals using MRT that had normal development and adult health, and in which mitochondria were predominantly derived from the donor egg. Pronuclear transfer (PNT) had been used in mice successfully since the 1980s, and mouse experiments allowed an examination of the degree of genetic distance between the donor's and mother's mitochondrial DNA (mtDNA) haplotypes that could be safely tolerated. Experiments in macaques using maternal spindle transfer (MST) were also successful; offspring lacked detectable maternal mtDNA and were healthy at the time of reporting. The 2016 panel also reviewed a limited amount of clinical data, most of which were unpublished. A

including many in Europe as well as the United States, where federal budget provisions currently prevent the FDA from considering any application for clinical use of HHGE (Kaiser, 2019). Any clinical use of HHGE in these countries would require changes to the relevant legislation.

All countries in which HHGE research and clinical applications may be pursued will need regulatory mechanisms to oversee use of HHGE and impose sanctions where appropriate, as well as a clearly communicated way for concerns about possible violations of regulations to be reported. Because HHGE would be deployed within an existing culture of IVF and ART clinics, it will be important to engage with this community on the issues posed by HHGE prior to any clinical uses. However, relying on professional conduct guidelines and self-regulation for an emerging and con-

report at the time of the review indicated that a baby had been born in Mexico following MRT. This suggested that MRT could be used to generate a healthy child but in the absence of full details (scientific and clinical) being made available, the panel was reluctant to rely on such data.

Crucial preclinical data came from human embryos that were generated using PNT and MST by different groups. The carryover of mtDNA from the mother's egg was usually very low, and embryos had developmental parameters that were comparable with control embryos (using measures such as fertilization rates and the proportion forming blastocysts). Transcript profiling suggested that the embryos generated through MRT and control embryos had comparable gene expression.

The expert panels prior to 2016 determined that it would be useful to examine the proportion of carried-over mtDNA in embryonic stem cells (ES cells) derived from such embryos in order to model the post-implantation embryo when mtDNA replication may be a factor. Three research groups independently observed that levels of carried-over mtDNA could elevate after extended ES cell passaging in vitro and come to predominate in around 20 percent of the cultures—a phenomenon called reversion. These data were important in the panel's cautious approach to introducing the technique into the clinic. It was recommended that only women with consistently high levels of pathogenic mtDNA in their oocytes—for whom PGT would likely not be successful—should be eligible for treatment when considering potential benefits and harms. Moreover, it was recommended that women be offered prenatal testing to assess mtDNA levels in the fetus to check for the possibility of reversion in vivo. Similarly, there had been, and still are, concerns about the possibility of functional mismatch between the mother's nuclear genome and the donor's mtDNA. At the time of review, there was no direct evidence for this. But the panel recommended that mtDNA haplotype matching be considered to mitigate any risk because what would be undertaken represented first-in-human uses of the technology and data were scarce.

troversial technology such as HHGE may be insufficient. At a minimum, laws or regulations incorporating penalties for any unauthorized use of HHGE should be considered in countries that do not currently have them.

Each country that considers the development of HHGE will end up drawing on the regulatory infrastructure and oversight authorities available under its laws and regulations. For a country such as the United Kingdom, the Human Fertilisation and Embryology Act could be further amended to permit HHGE, as it was in 2008 to enable the Authority to evaluate applications for MRT. If the U.S. government were to decide to permit clinical use of HHGE, it would also need to consider whether the FDA or other state and federal regulatory bodies need additional authorities to oversee the practice of assisted reproductive medicine, since HHGE would take place in ART clinics. Other countries may similarly need to wrestle with how HHGE could fit within or challenge national medical oversight systems and determine whether they need to create new oversight paradigms or whether existing oversight mechanisms could be modified to sufficiently address the oversight needs for HHGE.

REQUIREMENTS FOR NATIONAL OVERSIGHT SYSTEMS FOR HERITABLE HUMAN GENOME EDITING

Regardless of the details of the regulatory systems that a country may design for HHGE, national regulatory authorities or their equivalents would need to establish the specific criteria that must be met for any translational application of HHGE to proceed in their jurisdictions. To address the characteristics for responsible governance of HHGE identified above, all countries in which it is being considered would need to have mechanisms in place to oversee translational progress toward potential clinical use of HHGE, to prevent unpermitted uses and to sanction any misconduct. The issues that will need to be addressed through national systems wherever HHGE is proposed to be undertaken include:

- giving clear and unambiguous direction to researchers and clinicians about the legality of HHGE;
- ensuring that researchers and clinicians adhere to norms of responsible science, including relevant human rights and bioethics principles (see Box 5-3) and applicable guidelines, standards, and policies;
- providing transparency on any applications for HHGE under consideration;
- providing transparency to the world community of any intention to allow an approved clinical use of HHGE;
- creating clear processes and mechanisms for review, approval, and oversight of any initial human clinical uses of HHGE;

BOX 5-3
Adherence to Human Rights and to Bioethical Principles

Legally binding human rights and established human rights norms have been influential in framing the appropriate use of medical interventions.[a] Human rights language has long been conflated with biomedical ethics language, but their normative purpose and impact are different. Human rights are universal in their framing, although national translation in individual countries may incorporate cultural and religious differences. Human rights are legally actionable and belong to both individuals and collectivities. The language of policies and guidelines in biomedicine may use "rights" language, such as the "right of the child to an open future"; however, this is an ethical concept and not a legally recognized right of the child.

The need to develop governance approaches to encompass HHGE provides a potential opportunity to use and develop the content of internationally recognized human rights to influence future laws, policies, and regulatory responses around HHGE. However, the possibility of using human rights to frame, delimit, or expand concepts such as the freedom to conduct scientific research, the right of everyone to benefit from scientific advances, the right of children to the highest attainable standard of health, or even the rights of future generations has not yet been discussed by international bodies deliberating on HHGE. The feasibility of charging existing international and national supervisory or regulatory bodies with oversight of such rights in this specific context remains to be determined. This is an area that could be further explored as one foundation for future HHGE governance and is an approach being explored by the WHO Committee.

A foundational aspect for any use of HHGE is consideration of bioethics principles. Since the establishment of the Nuremberg Code in 1947, the field of biomedical research has benefited from international ethical guidance responsive to scientific developments (CIOMS, 2016; UNESCO, 2015; WMA, 2013). These norms, while self-regulatory in nature, have been influential in prospectively addressing areas of public concern, such as deliberate interventions in the human germline. Different reports and organizations present and interpret bioethical principles in slightly different ways, but they share many common features (NASEM, 2019a). In its 2017 report on human genome editing, for example, the U.S. National Academies identified seven principles to guide the development of governance for human genome editing that reflect this set of norms: promoting well-being, transparency, due care, responsible science, respect for persons, fairness, and transnational cooperation (NASEM, 2017). The Nuffield Council on Bioethics, in its report on genome editing and human reproduction, stated that two "overarching principles" should guide the use of HHGE interventions for them be ethically acceptable: they "should be used only where the procedure is carried out in a manner and for a purpose that is intended to secure the welfare of and is consistent with the welfare of a person who may be born as a consequence of treatment using those cells"; and they "should be permitted only in circumstances in which it cannot reasonably be expected to produce or exacerbate social division or the unmitigated marginalization or disadvantage of groups

continued

BOX 5-3 Continued

within society" (NCB, 2018, p. xvii). The WHO's Expert Advisory Committee on human genome editing has articulated six values, principles, and goals for good governance of these technologies: "a. Clarity, transparency and accountability; b. Responsible stewardship of resources; c. Inclusiveness, solidarity, and the common good; d. Fairness, non-discrimination, and social justice; e. Respect for the intrinsic dignity of the person; and f. Enforcement capacity" (WHO, 2020).

Because decision making about HHGE is ultimately a function of individual jurisdictions, countries should only permit the use of HHGE if they have engaged in thorough scrutiny of how any proposed application conforms with the human rights and ethical principles discussed above.

---

[a] Notable legal human rights instruments with regard to HHGE are the Convention on the Rights of the Child (1989), the Convention on Human Rights and Biomedicine (Oviedo) (1997), and the Convention on the Rights of Persons with Disabilities (2006).

- establishing mechanisms to circumscribe the clinical use of HHGE, including to limit and control any uses beyond the scope of a permitted indication; and
- being responsive to international scientific consensus regarding the current state of HHGE technologies, especially in areas of safety and proposed uses, with the goal of coordinating protocols and sharing data to the maximum extent possible.

THE NEED FOR A SYSTEM OF
GLOBAL COORDINATION AND COLLABORATION

While countries have decision-making authority concerning the research toward, or clinical use of HHGE, it is critical to also have international scientific and ethical cooperation on HHGE. A translational pathway for HHGE therefore requires governance systems that extend beyond those of individual countries to enable transparent discussion about any approved clinical uses of HHGE and the resulting outcomes. This is because:

- There is a collective interest of humanity in the use of a novel technology that can result in heritable changes to the human genome.
- The research and clinical communities developing these technologies are global, and the technologies have implications beyond national borders.

- Citizens from different countries seeking access to HHGE will travel to countries where it becomes available.
- Any initial uses of HHGE following the pathways described in this report would involve a small number of people, and it would be important to collect and compare information across national boundaries in order to more fully understand first-in-human safety and efficacy data and to promote common approaches.

With respect to both biomedical research and clinical practice, in general, countries have framed licensing powers and accompanying professional duties within their legislation and regulations on health and their health systems, through the creation of statutory oversight bodies, or, more rarely, via legislation on specific sectors or technologies. The approach varies from country to country or it is defined in regional alliances. Any proposed mechanism for international governance of HHGE will need to provide for at least three functions:

1. An international scientific advisory panel to provide ongoing technical assessment and evaluation of developments in the science and technologies on which HHGE depends and to make recommendations about their suitability and readiness for particular clinical uses.
2. An international body for evaluating and making recommendations on crossing major thresholds associated with the clinical use of HHGE, based on consideration of a wide range of societal and scientific perspectives. In the current context, a threshold represents a boundary that distinguishes a currently accepted use from another that is not currently accepted. Before crossing any threshold, it will be important for the global community to assess not only progress in scientific research but also what additional ethical and societal concerns the circumstances of particular uses could raise, as well as any results, successes, or concerns that had been observed from any human uses of HHGE that had been conducted thus far.
3. An international mechanism by which individuals or organizations in one country can bring forward technical or ethical concerns arising from HHGE work conducted in their own country or in another country.

These necessary functions are explored below.

An International Scientific Advisory Panel to Monitor and Assess Relevant Scientific and Clinical Developments

As emphasized throughout the report, before any country should consider approving the clinical use of HHGE, further technical developments are essential. There is therefore a need for the ongoing technical assessment and evaluation of developments in the science and technologies on which HHGE depends, as well as making recommendations about their suitability and readiness for particular clinical uses. Multiple gaps in the ability to fully characterize such genome editing or assess its effects make it premature to use any HHGE approaches at the time of this writing, and articulating the essential characteristics of a translational pathway does not mean that a country should necessarily permit even initial clinical uses.

There is, therefore, a need for an international advisory body to regularly review the latest scientific evidence and to evaluate its potential impact on the feasibility of HHGE. The necessary functions of such scientific review include:

- assessing or making recommendations on further research developments that would be required to reach technical or translational milestones as research on HHGE progresses;
- providing information to national regulatory authorities or their equivalents to inform their own assessment and oversight efforts;
- facilitating coordination or standardization of study designs to promote the ability to compare and pool data across studies and trans-nationally;
- advising on specific measures to be used as part of the long-term follow-up of any children born following HHGE; and
- reviewing data on clinical outcomes from any regulated uses of HHGE and advising on the risks, benefits, and uncertainties of possible further applications.

There are existing international activities that play a valuable role in contributing to the technical assessment of the science and technologies underlying HHGE. The two international summits on human genome editing convened by various scientific academies (including the U.S. National Academies of Sciences and Medicine, the U.K.'s Royal Society, the Chinese Academy and the Hong Kong Academy, and others) have brought together the scientific community for scientific presentations relevant to HHGE. A third summit is planned for 2021.

Professional societies in science or medicine can also play a role in scientific review and standards development. In stem cell research, the International Society for Stem Cell Research has an ongoing mechanism for the creation and revision of guidelines (ISSCR, 2016) for research and clinical practice in the stem cell field. In the ART field, the U.S. Society for Assisted Reproductive Technology (SART) provides access to data from IVF clinics for research and comparison and is developing a standardized document for informed consent in collaboration with the American Association of Law Schools.[3]

However, these activities are largely informal and ad hoc. The examples demonstrate that although existing structures and processes can fulfill some of the functions necessary, none do or can perform the collection of functions recommended for ongoing technical assessment and evaluation of the technologies foundational to HHGE.

For this reason, the Commission recommends the creation of an International Scientific Advisory Panel (ISAP) that would provide regular scientific and technical assessments as part of the international governance efforts for HHGE described above (see Figure 5-1). An ISAP would need the endorsement of national governments to have the standing and influence required to perform these functions. It would also need to be flexible given the potential for rapid advances in areas of science that contribute to the feasibility of HHGE. The panel would need to convene regularly in person or virtually, likely at least once per year, with additional meetings and discussions as needed. To be most effective, such a panel would need to have diverse, multidisciplinary membership and include independent experts who can assess scientific evidence of safety and efficacy of both genome editing and associated ARTs. It should include international experts from multiple disciplines including genetics, genome editing, reproductive medicine, pediatric and adult medicine, bioethics, law, and other fields. This combination is similar to that for Data Safety and Monitoring Boards or Data Monitoring Committees for large, often multi-site, clinical trials, which seek to ensure relevant expertise in clinical specialty areas, clinical trial methodologies and analysis, biostatistics, and often in the ethics of design, conduct, and interpretation of clinical trials.[4] Because the panel would be assessing evidence that could be used to support progress toward initial use of HHGE for serious monogenic diseases, the panel would also greatly benefit from including representatives of the public, such as members of genetic disease and disability communities.

---

[3] See www.sart.org. Recent SART clinic reports are available at https://www.sartcorsonline.com/rptCSR_PublicMultYear.aspx?reportingYear=2017 for 2017, and preliminary 2018 data at https://www.sartcorsonline.com/rptCSR_PublicMultYear.aspx?reportingYear=2018.

[4] See FDA guidance at https://www.fda.gov/regulatory-information/search-fda-guidance-documents/establishment-and-operation-clinical-trial-data-monitoring-committees; and NIH guidelines at https://www.nidcr.nih.gov/research/human-subjects-research/toolkit-and-education-materials/interventional-studies/data-and-safety-monitoring-board-guidelines.

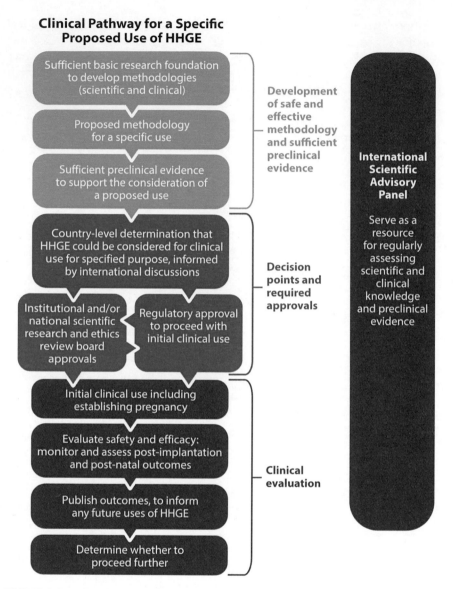

FIGURE 5-1 An ISAP would provide regular, independent assessments along the clinical translational pathway, as envisioned in Chapter 4, toward HHGE for certain circumstances of serious monogenic diseases. These assessments would include reviewing advances in preclinical research, providing advice on whether sufficient methodologies to support evaluating a proposed use had been developed, informing the deliberations of a country's own advisory or regulatory bodies if requested, and analyzing the outcomes of any permitted clinical uses of HHGE.

Existing national and international networks could be drawn on to identify members who could be nominated to such a panel. For example, national academies of sciences and medicine, the global network of science academies (the InterAcademy Partnership), national and international professional societies in relevant areas, genetic disease and disability communities, and scientific, medical, and technical experts in relevant government ministries might all serve to identify colleagues who are leaders in their disciplines and could bring the expertise and cooperative spirit required to this task. National and international discussions would be needed to agree on the panel's terms of reference, its convener, and how its activities would be funded.

The Commission is not wed to any particular body or organization for establishing an ISAP but emphasizes its recommendation that any translational pathway toward HHGE requires establishing a systematic and rigorous way to fulfill the five functions described above in order to enable independent expert review of scientific and clinical evidence to inform national and international governance.

Advances in areas of science relevant to HHGE, as well as in the practice of IVF and PGT, will have implications for whether the translational pathway criteria specified in Chapter 4 can be met. As this pathway was developed for the very first possible uses of HHGE considering the current state of science, it will be important to be open to scientific developments that could alter the methodologies employed to meet the requirements. It will also be important to assess evidence gained from further basic research and preclinical testing and from any future initial human uses.

The Commission strongly believes that successfully carrying out these functions requires more than the current informal and ad hoc systems.

International Body for Evaluating and Making Recommendations before Crossing Heritable Human Genome Editing Thresholds

This report has categorized possible clinical uses of HHGE according to the assessment of the potential harms and benefits they present, with a focus on initial clinical uses. However, decisions about whether to allow HHGE, and, if so, for what purposes, should be based on a wider set of considerations than just scientific assessments. Initial human use of HHGE beyond preclinical development represents a decision that should be based on science, ethics, and societal implications. It will be important for countries to engage in discussions about when, if ever, it is acceptable to move forward with HHGE within their countries and, if so, where to set thresholds on allowable uses. Subsequent decisions about whether to cross additional thresholds to allow further uses of HHGE will similarly require transparent international discussions convened by an institution responsible for ensuring that these discussions are held regularly and that they engage a diversity of viewpoints (see Figure 5-2).

There is already a range of international bodies whose responsibilities include convening international discussions on the development and regulation of medical technologies. Organizations such as WHO, the Organisation for Economic Cooperation and Development (OECD), and

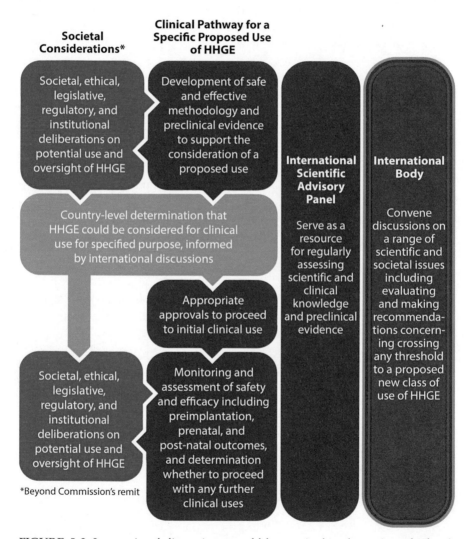

FIGURE 5-2 International discussions would be required to determine whether it would be possible to cross significant thresholds and describe translational pathways for potential uses of HHGE.

the United Nations Educational, Scientific and Cultural Organization (UNESCO), for example, all have the requisite experience that could enable an inclusive and transparent debate about whether and how to proceed with HHGE. Other organizations could also be selected for this purpose.

Regardless of where housed or how structured, this international body would need a wide range of perspectives, including from (i) stakeholder communities that could be affected by future uses of HHGE, such as members of disability and disease communities; (ii) scientific fields, including medicine and social sciences; and (iii) law, ethics, and regulation. This should include experts from countries where there are communities that have increased incidence of genetic disease due to factors such as founder mutations or high rates of consanguineous union. As with the current Commission, assessments from this process would inform and be advisory to national and international decision making.

If initial clinical uses of HHGE were ever permitted, those uses would only be considered in a carefully prescribed set of circumstances and would likely entail only a small number (on the order of 10–20) of cases. Assuming analysis of the outcomes of any initial uses did not raise further concerns about the safety and efficacy of HHGE, it might be deemed appropriate to consider uses in circumstances beyond those initially envisaged by this Commission. Before progressing beyond those initial cases toward any further clinical uses of HHGE, it would be important for the global community to pause and reassess not only the state of the science but also what additional ethical and societal concerns new circumstances of uses could raise. New classes of use may or may not precisely align with the six categories defined in Chapter 3. Making recommendations on whether it is appropriate to cross subsequent thresholds in the use of HHGE would be a key role for an international body with responsibility for convening the international debate on HHGE.

Potential uses of HHGE beyond the circumstances set out by this Commission open the door to impacting reproductive options for a significantly larger group of people. Making HHGE available to couples in Category B beyond the narrow circumstances described in Chapter 3 would represent a significant expansion in the possible scope of this technology. As a result, a respected body would be needed that can assess whether it is feasible to envision new responsible translational pathways and what these pathways should entail.

This process should be complemented by other efforts by civil society to promote international cooperation on approaches to responsible development of medical technologies. For example, the Global Observatory on Genome Editing is being set up to foster international, interdisciplinary discussions on genome editing (Hurlbut et al., 2018). Similarly, the Association for Responsible Research and Innovation in Genome Editing (ARRIGE) was launched in 2018 to promote global governance of genome editing.[5] Both ARRIGE and the Global Observatory promote cross-sector discussions of whether genome editing technologies should be used and, if so, for what purposes.

There are also international processes that focus more on promoting responsible scientific conduct, for example, the Good Clinical Practice guidance for clinical trials developed by the International Council for Harmonisation of Technical Requirements for Pharmaceuticals for Human Use (ICH), whose members and observers include national regulatory agencies, industry, and international organizations.[6] The ICH develops its guidelines through a process that includes formation of an expert working group to draft a technical document on an issue, followed by development by regulatory members of a draft guideline. The draft guideline undergoes a process of consultation and revision before being adopted by ICH. Although governance of HHGE would require engaging a broader and more diverse community than encompassed by ICH, this step-wise process encourages input and buy-in from the represented stakeholders. The launch by WHO in 2019 of initial phases of a global registry for genome editing clinical trials also represents an important step in the ability to monitor advances in HHGE and to maintain awareness of actions being taken within national jurisdictions.[7]

A Mechanism to Bring Forward Concerns about Research or Clinical Use of Heritable Human Genome Editing

After the announcement in 2018 that children had been born in China following the use of HHGE, an important question posed was how individuals who may have known about the work being conducted could have raised concerns, particularly if they were in one country and the investigator and the research being undertaken were in another. The Commission is not aware of a precise precedent for such an international mechanism that is accessible to anyone who would like to raise a concern.

Future governance of HHGE requires an international mechanism for individuals and groups to raise concerns about possible violations of regulations or activities related to any clinical practice of HHGE in jurisdictions without regulations. There should be a highly visible, easily discoverable

---

[5] See https://www.arrige.org.

[6] See https://www.ich.org.

[7] Information on the registry is available at https://www.who.int/health-topics/ethics/human-genome-editing-registry/. The registry collects information on clinical trials using somatic genome editing as well as any clinical trials that would be conducted using HHGE.

entity to which people everywhere may direct their concerns about activity in any country. In developing this mechanism, it will be important to keep in mind that raising concerns about scientific or clinical practices can have personal and professional ramifications for the person making the complaint. It is therefore important to maintain anonymity for anyone using this service. Similarly, details of a complaint should not be made public without prior investigation to protect individuals, institutions, and businesses from false accusations. Such investigations would be the responsibility of national regulatory authorities where available. These authorities would be informed by the international mechanism that a complaint had been made against someone within their jurisdiction.

Although there is no exact precedent, there are relevant examples that can inform the design of such a mechanism. The World Anti-Doping Agency (WADA) has a means by which anyone can report an "Alleged Anti-Doping Rule Violation or any act or omission that could undermine the fight against doping."[8] Some research funders have also developed mechanisms to facilitate the investigation of complaints made against researchers they fund.

CONCLUSION AND RECOMMENDATIONS

The pursuit of a translational pathway toward the clinical use of HHGE would represent the controlled alteration of a human embryonic genome using genome-editing tools, offered as part of an assisted reproduction intervention. All countries pursuing research on or considering the use of HHGE will need to establish oversight systems for this technology, even though national regulatory frameworks for HHGE will differ in their structures and approaches. The governance structures needed for HHGE will also require new models of international coordination. Complex scientific and clinical information will need to be assessed to identify whether the criteria for clinically evaluating a proposed use of HHGE can be met and to incorporate any resulting outcomes into future discussions and decision making. Achieving national and international coordination will pose challenges. But this is exactly why it will be critical to create robust processes by which there can be appropriate and transparent shared responsibility for moving HHGE forward thoughtfully and cautiously, only if there is clear scientific consensus to continue and only if a given country decides to permit its use.

The Commission recommends the following actions as part of this process.

---

[8] See https://speakup.wada-ama.org/FrontPages/Default.aspx.

Essential Elements of Oversight Systems for
Heritable Human Genome Editing

Important national and international governance mechanisms should be established before any clinical use of HHGE.

> **Recommendation 8:** Any country in which the clinical use of heritable human genome editing (HHGE) is being considered should have mechanisms and competent regulatory bodies to ensure that all of the following conditions are met:
>   - individuals conducting HHGE-related activities, and their oversight bodies, adhere to established principles of human rights, bioethics, and global governance;
>   - the clinical pathway for HHGE incorporates best practices from related technologies such as mitochondrial replacement techniques, preimplantation genetic testing, and somatic genome editing;
>   - decision making is informed by findings from independent international assessments of progress in scientific research and the safety and efficacy of HHGE, which indicate that the technologies are advanced to a point that they could be considered for clinical use;
>   - prospective review of the science and ethics of any application to use HHGE is diligently performed by an appropriate body or process, with decisions made on a case-by-case basis;
>   - notice of proposed applications of HHGE being considered is provided by an appropriate body;
>   - details of approved applications (including genetic condition, laboratory procedures, laboratory or clinic where this will be done, and national bodies providing oversight) are made publicly accessible, while protecting family identities;
>   - detailed procedures and outcomes are published in peer-reviewed journals to provide dissemination of knowledge that will advance the field;
>   - the norms of responsible scientific conduct by individual investigators and laboratories are enforced;
>   - researchers and clinicians show leadership by organizing and participating in open international discussions on the coordination and sharing of results of relevant scientific, clinical, ethical, and societal developments impacting the assessment of HHGE's safety, efficacy, long-term monitoring, and societal acceptability;

- practice guidelines, standards, and policies for clinical uses of HHGE are created and adopted prior to offering clinical use of HHGE; and
- reports of deviation from established guidelines are received and reviewed, and sanctions are imposed where appropriate.

Recommendation 9: An International Scientific Advisory Panel (ISAP) should be established with clear roles and responsibilities before any clinical use of heritable human genome editing (HHGE). The ISAP should have a diverse, multidisciplinary membership and should include independent experts who can assess scientific evidence of safety and efficacy of both genome editing and associated assisted reproductive technologies. The ISAP should:

- provide regular updates on advances in, and the evaluation of, the technologies that HHGE would depend on and recommend further research developments that would be required to reach technical or translational milestones;
- assess whether preclinical requirements have been met for any circumstances in which HHGE may be considered for clinical use;
- review data on clinical outcomes from any regulated uses of HHGE and advise on the scientific and clinical risks and potential benefits of possible further applications; and
- provide input and advice on any responsible translational pathway to the international body described in Recommendation 10, as well as at the request of national regulators.

Recommendation 10: In order to proceed with applications of heritable human genome editing (HHGE) that go beyond the translational pathway defined for initial classes of use of HHGE, an international body with appropriate standing and diverse expertise and experience should evaluate and make recommendations concerning any proposed new class of use. This international body should:

- clearly define each proposed new class of use and its limitations;
- enable and convene ongoing transparent discussions on the societal issues surrounding the new class of use;
- make recommendations concerning whether it could be appropriate to cross the threshold of permitting the new class of use; and

- provide a responsible translational pathway for the new class of use.

Recommendation 11: An international mechanism should be established by which concerns about research or conduct of heritable human genome editing that deviates from established guidelines or recommended standards can be received, transmitted to relevant national authorities, and publicly disclosed.

References

Aach, J., J. Lunshof, E. Iyer, and G. M. Church. 2017. Addressing the ethical issues raised by synthetic human entities with embryo-like features. *eLife* 6:e20674. doi:10.7554/eLife.20674.

Abou-El-Enein, M., T. Cathomen, Z. Ivics, C. H. June, M. Renner, C. K. Schneider, and G. Bauer. 2017. Human genome editing in the clinic: New challenges in regulatory benefit-risk assessment. *Cell Stem Cell* 21(4):427–430. doi:10.1016/j.stem.2017.09.007.

Acuna-Hidalgo, R., J. A. Veltman, and A. Hoischen. 2016. New insights into the generation and role of de novo mutations in health and disease. *Genome Biology* 17(1):241. doi:10.1186/s13059-016-1110-1.

Adashi, E. Y., and I. G. Cohen. 2019. Heritable genome editing: Is a moratorium needed? *JAMA* 322(2):104–105. doi:10.1001/jama.2019.8977.

Adashi, E. Y., I. G. Cohen, J. H. Hanna, A. M. Surani, and K. Hayashi. 2019. Stem cell–derived human gametes: The public engagement imperative. *Trends in Molecular Medicine* 25(3):165–167. doi:10.1016/j.molmed.2019.01.005.

Adikusuma, F., S. Piltz, M. A. Corbett, M. Turvey, S. R. McColl, K. J. Helbig, M. R. Beard, J. Hughes, R. T. Pomerantz, and P. Q. Thomas. 2018. Large deletions induced by Cas9 cleavage. *Nature* 560:E8–E9. doi:10.1038/s41586-018-0380-z.

Alanis-Lobato, G., J. Zohren, A. McCarthy, N. M. E. Fogarty, N. Kubikova, E. Hardman, M. Greco, D. Wells, J. M. A. Turner, and K. K Niakan. 2020. Frequent loss-of-heterozygosity in CRISPR-Cas9-edited early human embryos. *bioRxiv* 2020.06.05.135913. doi:10.1101/2020.06.05.135913.

Altarescu, G., B. Brooks, T. Eldar-Geva, E. J. Margalioth, A. Singer, E. Levy-Lahad, and P. Renbaum. 2008. Polar body-based preimplantation genetic diagnosis for N-acetylglutamate synthase deficiency. *Fetal Diagnosis and Therapy* 24(3):170–176. doi:10.1159/000151333.

ANM (Académie Nationale de Médecine). 2016. Genome editing of human germline cells and embryos. Paris, France. http://www.academie-medecine.fr/wp-content/uploads/2016/05/report-genome-editing-ANM-2.pdf.

Anzalone, A. V., P. B. Randolph, J. R. Davis, A. A. Sousa, L. W. Koblan, J. M. Levy, P. J. Chen, C. Wilson, G. A. Newby, A. Raguram, and D. R. Liu. 2019. Search-and-replace genome editing without double-strand breaks or donor DNA. *Nature* 576:149–157. doi:10.1038/s41586-019-1711-4.

Archer, N. M., N. Peterson, M. A. Clark, C. O. Buckee, L. M. Childs, and M. T. Duraising. 2018. Resistance to *Plasmodium falciparum* in sickle cell trait erythrocytes is driven by oxygen-dependent growth inhibition. *PNAS* 115(28):7350–7355. https://doi.org/10.1073/pnas.1804388115.

ASGCT (American Society of Gene and Cell Therapy). 2019. Letter to HHS Secretary Azar. https://www.asgct.org/global/documents/clinical-germline-gene-editing-letter.aspx.

Bailey, S. R., and M. V. Maus. 2019. Gene editing for immune cell therapies. *Nature Biotechnology* 37:1425–1434. doi:10.1038/s41587-019-0137-8.

Bay, B., H. J. Ingerslev, J. G. Lemmen, B. Degn, I. A. Rasmussen, and U. S. Kesmodel. 2016. Preimplantation genetic diagnosis: A national multicenter obstetric and neonatal follow-up study. *Fertility and Sterility* 106(6):1363–1369. doi:10.1016/j.fertnstert.2016.07.1092.

Bayefsky, M. 2018. Who should regulate preimplantation genetic diagnosis in the United States? *AMA Journal of Ethics* 20(12):E1160–1167. doi: 10.1001/amajethics.2018.1160.

Bayefsky, M. J. 2016. Comparative preimplantation genetic diagnosis policy in Europe and the USA and its implications for reproductive tourism. *Reproductive Biomedicine and Society Online* 3:41–47. doi:10.1016/j.rbms.2017.01.001.

Ben Khelifa, M., R. Zouari, R. Harbuz, L. Halouani, C. Arnoult, J. Lunardi, and P. F. Ray. 2011. A new AURKC mutation causing macrozoospermia: Implications for human spermatogenesis and clinical diagnosis. *Molecular Human Reproduction* 17(12):762–768. doi:10.1093/molehr/gar050.

Bender, W., M. Akam, F. Karch, P. A. Beachy, M. Peifer, P. Spierer, E. B. Lewis, and D. S. Hogness. 1983. Molecular genetics of the bithorax complex in *Drosophila melanogaster*. *Science* 221(4605):23–29. doi:10.1126/science.221.4605.23.

Berntsen, S., V. Soderstrom-Antilla, U. B. Wennerholm, H. Laivuori, A. Loft, N. B. Oldereid, L. B. Romundstad, C. Bergh, and A. Pinborg. 2019. The health of children conceived by ART: "The chicken or the egg?" *Human Reproduction Update* 25(2):137–158. doi:10.1093/humupd/dmz001.

Bibikova, M., K. Beumer, J. K. Trautman, and D. Carroll. 2003. Enhancing gene targeting with designed zinc finger nucleases. *Science* 300(5620):764. doi:10.1126/science.1079512.

Bioethics Advisory Committee, Singapore. 2018. *Ethical, Legal and Social Issues Arising From Mitochondrial Genome Replacement Therapy: A Consultation Paper*. https://www.bioethics-singapore.gov.sg/files/publications/consultation-papers/mitochondrial-genome-replacement-tech.pdf.

Blair, D. R., C. S. Lyttle, J. M. Mortensen, C. F. Bearden, A. B. Jensen, H. Khiabanian, R. Melamed, R. Rabadan, E. V. Bernstam, S. Brunak, L. J. Jensen, D. Nicolae, N. H. Shah, R. L. Grossman, N. J. Cox, K. P. White, and A. Rzhetsky. 2013. A nondegenerate code of deleterious variants in Mendelian loci contributes to complex disease risk. *Cell* 155(1):70–80. doi:10.1016/j.cell.2013.08.030.

Bosley, K. S., M. Botchan, A. L. Bredenoord, D. Carroll, R. A. Charo, E. Charpentier, R. Cohen, J. Corn, J. Doudna, G. Feng, H. T. Greely, R. Isasi, W. Ji, J. S. Kim, B. Knoppers, E. Lanphier, J. Li, R. Lovell-Badge, G. S. Martin, J. Moreno, L. Naldini, M. Pera, A. C. F. Perry, J. C. Venter, F. Zhang, and Q. Zhou. 2015. CRISPR germline engineering: The community speaks. *Nature Biotechnology* 33(5):478–486. doi:10.1038/nbt.3227.

Botstein, D., R. L. White, M. Skolnick, and R. W. Davis. 1980. Construction of a genetic linkage map in man using restriction fragment length polymorphisms. *American Journal of Human Genetics* 32(3):314–331.

Braude, P. 2019. Assisted reproduction techniques for avoiding inherited diseases: Practical aspects of PGD and results. Presentation to the International Commission on Clinical Use of Human Germline Genome Editing, November 19.

Bredenoord, A. L., and I. Hyun. 2017. Ethics of stem cell–derived gametes made in a dish: Fertility for everyone? *EMBO Molecular Medicine* 9:396–398. doi:10.15252/emmm.201607291.

Brokowski, C. 2018. Do CRISPR germline ethics statements cut it? *The CRISPR Journal* 1(2):115–125. doi:10.1089/crispr.2017.0024.

Brown, S. D. M., and H. V. Lad. 2019. The dark genome and pleiotropy: Challenges for precision medicine. *Mammalian Genome* 30(7–8):212–16. doi:10.1007/s00335-019-09813-4.

Brzeziańska, E., D. Domańska, and A. Jegier. 2014. Gene doping in sport—perspectives and risks. *Biology of Sport* 31(4):251–259. doi:10.5604/20831862.112093.

Burnham-Marusich, A. R., C. O. Ezeanolue, M. C. Obiefune, W. Yang, A. Osuji, A. G. Ogidi, A. T. Hunt, D. Patel, and E. E. Ezeanolue. 2016. Prevalence of sickle cell trait and reliability of self-reported status among expectant parents in Nigeria: Implications for targeted newborn screening. *Public Health Genomics* 19(5):298–306. doi: 10.1159/000448914.

Cacheiro, P., V. Muñoz-Fuentes, S. A. Murray, M. E. Dickinson, M. Bucan, L. M. J. Nutter, K. A. Peterson, H. Haselimashhadi, A. M. Flenniken, H. Morgan, H. Westerberg, T. Konopka, C. Hsu, A. Christiansen, D. G. Lanza, A. L. Beaudet, J. D. Heaney, H. Fuchs, V. Gailus-Durner, T. Sorg, J. Prochazka, V. Novosadova, C. J. Lelliott, H. Wardle-Jones, S. Wells, L. Teboul, H. Cater, M. Stewart, T. Hough, W. Wurst, R. Sedlacek, D. J. Adams, J. R. Seavitt, G. Tocchini-Valentini, F. Mammano, R. E. Braun, C. McKerlie, Y. Herault, M. Hrabě de Angelis, A. Mallon, K. C. K. Lloyd, S. D. M. Brown, H. Parkinson, T. F. Meehan, D. Smedley, The Genomics England Research Consortium, and The International Mouse Phenotyping Consortium. 2020. Human and mouse essentiality screens as a resource for disease gene discovery. *Nature Communications* 11:655. doi: 10.1038/s41467-020-14284-2.

Capecchi, M. 2005. Gene targeting in mice: functional analysis of the mammalian genome for the twenty-first century. *Nature Reviews Genetics* 6:507–512. doi:10.1038/nrg1619.

Cavaliere, G. 2017. A 14-day limit for bioethics: The debate over human embryo research. *BMC Medical Ethics* 18:38. doi:10.1186/s12910-017-0198-5.

CEST (Commission de l'Éthique en Science et en Technologie). 2019. *Genetically modified babies: Ethical issues raised by the genetic modification of germ cells and embryos.* Québec City, Québec. https://www.ethique.gouv.qc.ca/media/1038/cest_modif_gene_resume_an_acc.pdf.

Chen, D., N. Sun, L. Hou, R. Kim, J. Faith, M. Aslanyan, Y. Tao, Y. Zheng, J. Fu, W. Liu, M. Kellis, and A. Clark. 2019. Human primordial germ cells are specified from lineage-primed progenitors. *Cell Reports* 29(13):4568–4582. doi:10.1016/j.celrep.2019.11.083.

Chen, J. S., Y. S. Dagdas, B. P. Kleinstiver, M. M. Welch, A. A. Sousa, L. B. Harrington, S. H. Sternberg, J. K. Joung, A. Yildiz, and J. A Doudna. 2017. Enhanced proofreading governs CRISPR-Cas9 targeting accuracy. *Nature* 550(7676): 407–410. doi.org/10.1038/nature24268.

Chen, Y., Y. Zheng, Y. Kang, W. Yang, Y. Niu, X. Guo, Z. Tu, C. Si, H. Wang, R. Xing, X. Pu, S. H. Yang, S. Li, W. Ji, and X. J. Li. 2015. Functional disruption of the dystrophin gene in rhesus monkey using CRISPR/Cas9. *Human Molecular Genetics* 24(13):3764–3774. doi:10.1093/hmg/ddv120.

CIOMS (Council for International Organizations of Medical Sciences). 2016. *International ethical guidelines for health-related research involving humans.* Geneva, Switzerland. https://cioms.ch/wp-content/uploads/2017/01/WEB-CIOMS-EthicalGuidelines.pdf.

Cioppi, F., E. Casamonti, and C. Krausz. 2019. Age-dependent de novo mutations during sper-matogenesis and their consequences. *Advances in Experimental Medicine and Biology* 1166:29–46. doi:10.1007/978-3-030-21664-1_2.

Claussnitzer, M., J. H. Cho, R. Collins, N. J. Cox, E. T. Dermitzakis, M. E. Hurles, S. Kathiresan, E. E. Kenny, C. M. Lindgren, D. G. MacArthur, K. N. North, S. E. Plon, H. L. Rehm, N. Risch, C. N. Rotimi, J. Shendure, N. Soranzo, and M. I. McCarthy. 2020. A brief history of human disease genetics. *Nature* 577:179–189. doi.org/10.1038/s41586-019-1879-7

Cohen, I. G., E. Y. Adashi, S. Gerke, C. Palacios-González, and V. Ravitsky. 2020. The regulation of mitochondrial replacement techniques around the world. *Annual Review of Genomics and Human Genetics* 21:1. doi:10.1146/annurev-genom-111119-101815.

Cohen, J. 2019a. Did CRISPR help—or harm—the first-ever gene-edited babies? https://www.sciencemag.org/news/2019/08/did-crispr-help-or-harm-first-ever-gene-edited-babies.

Cohen, J. 2019b. Inside the circle of trust. *Science* 365(6452):430–437. https://science.sciencemag.org/content/365/6452/430. doi:10.1126/science.365.6452.430.

Cuchel, M., E. Bruckert, H. N. Ginsberg, F. J. Raal, R. D. Santos, R. A. Hegele, J. A. Kuivenhoven, B. G. Nordestgaard, O. S. Descamps, E. Steinhagen-Thiessen, A. Tybjærg-Hansen, G. F. Watts, M. Averna, C. Boileau, J. Borén, A. L. Catapano, J. C. Defesche, G. K. Hovingh, S. E. Humphries, P. T. Kovanen, L. Masana, P. Pajukanta, K. G. Parhofer, K. K. Ray, A. F. Stalenhoef, E. Stroes, M. R. Taskinen, A. Wiegman, O. Wiklund, M. J. Chapman, and the European Atherosclerosis Society Consensus Panel on Familial Hypercholesterolaemia. 2014. Homozygous familial hypercholesterolaemia: New insights and guidance for clini-cians to improve detection and clinical management; A position paper from the Consensus Panel on Familial Hypercholesterolaemia of the European Atherosclerosis Society. *European Heart Journal* 35(32):2146–2157. doi:10.1093/eurheartj/ehu274.

Cyranoski, D. 2019. The CRISPR-baby scandal: What's next for human gene-editing. *Nature* 566:440. doi:10.1038/d41586-019-00673-1.

De Geyter, C., C. Calhaz-Jorge, M. S. Kupka, C. Wyns, E. Mocanu, T. Motrenko, G. Scaravelli, J. Smeenk, S. Vidakovic, V. Goossens, and the European IVF-monitoring Consortium (EIM) for the European Society of Human Reproduction and Embryology. 2020. ART in Europe, 2020: Results generated from European registries by ESHRE. Tables SV, SVI, and SVII. *Human Reproduction Open* 2020(1):hoz038. doi:10.1093/hropen/hoz038.

De Rycke, M., V. Goossens, G. Kokkali, M. Meijer-Hoogeveen, E. Coonen, and C. Moutou. 2017. ESHRE PGD Consortium data collection XIV–XV: Cycles from January 2011 to December 2012 with pregnancy follow-up to October 2013. *Human Reproduction* 32(10):1974–1994. doi:10.1093/humrep/dex265.

De Sanctis, V., C. Kattamis, D. Canatan, A. T. Soliman, H. Elsedfy, M. Karimi, S. Daar, Y. Wali, M. Yassin, N. Soliman, P. Sobti, S. Al Jaouni, M. El Kholy, B. Fiscina, and M. Angastiniotis. 2017. β-Thalassemia distribution in the Old World: An ancient disease seen from a historical standpoint. *Mediterranean Journal of Hematology and Infectious Diseases* 9(1):e2017018. doi:10.4084/MJHID.2017.018.

Delahaye, F., C. Do, Y. Kong, R. Ashkar, M. Salas, B. Tycko, R. Wapner, and F. Hughes. 2018. Genetic variants influence on the placenta regulatory landscape. *PLOS Genetics* 14(11):e1007785. doi.org/10.1371/journal.pgen.1007785.

Deltas, C. 2018. Digenic inheritance and genetic modifiers. *Clinical Genetics* 93(3):429–438. doi:10.1111/cge.13150.

Doetschman, T., R. G. Gregg, N. Maeda, M. L. Hooper, D. W. Melton, S. Thompson, and O. Smithies. 1987. Targeted correction of a mutant HPRT gene in mouse embryonic stem cells. *Nature* 330:576–578. doi:10.1038/330576a0.

Doman, J. L., A. Raguram, G. A. Newby, and D. R. Liu. 2020. Evaluation and minimization of Cas9-independent off-target DNA editing by cytosine base editors. *Nature Biotechnology* 38:620–628. doi:10.1038/s41587-020-0414-6.

Doudna, J. A., and E. Charpentier. 2014. Genome editing: The new frontier of genome engineering with CRISPR-Cas9. *Science* 346(6213):1258096. doi:10.1126/science.1258096.

Doyon, Y., T. D. Vo, M. C. Mendel, S. G. Greenberg, J. Wang, D. F. Xia, J. C. Miller, F. D. Urnov, P. D. Gregory, and M. C. Holmes. 2011. Enhancing zinc-finger-nuclease activity with improved obligate heterodimeric architectures. *Nature Methods* 8:74–79. doi:10.1038/nmeth.1539.

Eckersley-Maslin, M. A., C. Alda-Catalinas, and W. Reik. 2018. Dynamics of the epigenetic landscape during the maternal-to-zygotic transition. *Nature Reviews Molecular Cell Biology* 19:436–450. doi.org/10.1038/s41580-018-0008-z.

EGE (European Group on Ethics in Science and New Technologies). 2016. Statement on gene editing. Brussels, Belgium. https://ec.europa.eu/research/ege/pdf/gene_editing_ege_statement.pdf.

Eggermann, T., G. Perez de Nanclares, E. R. Maher, I. K. Temple, Z. Tümer, D. Monk, D. J. Mackay, K. Grønskov, A. Riccio, A. Linglart, and I. Netchine. 2015. Imprinting disorders: A group of congenital disorders with overlapping patterns of molecular changes affecting imprinted loci. *Clinical Epigenetics* 7:123. doi:10.1186/s13148-015-0143-8.

Egli, D., M. V. Zuccaro, M. Kosicki, G. M. Church, A. Bradley, and M. Jasin. 2018. Inter-homologue repair in fertilized human eggs? *Nature* 560:E5–E7. doi:10.1038/s41586-018-0379-5.

European Working Group on Cystic Fibrosis Genetics. 1990. Gradient of distribution in Europe of the major CF mutation and of its associated haplotype. *Human Genetics* 85:436–445.

Evans, J. H. 2002. *Playing God? Human genetic engineering and the rationalization of public bioethical debate.* Chicago, IL: University of Chicago Press.

Evans, S. J., I. Douglas, M. D. Rawlins, N. S. Wexler, S. J. Tabrizi, and L. Smeeth. 2013. Prevalence of adult Huntington's disease in the UK based on diagnoses recorded in general practice records. *Journal of Neurology, Neurosurgery, and Psychiatry* 84(10):1156–1160. doi:10.1136/jnnp-2012-304636.

Farrell, P. M. 2008. The prevalence of cystic fibrosis in the European Union. *Journal of Cystic Fibrosis* 7:450–453.

FEAM (Federation of European Academies of Medicine). 2017. *Human genome editing in the EU. Report of a workshop held on April 28, 2016 at the French Academy of Medicine.* Brussels, Belgium. https://www.interacademies.org/publication/feam-human-genome-editing-eu.

Fletcher, J. 1971. Ethical aspects of genetic controls: Designed genetic changes in man. *New England Journal of Medicine* 285(14):776–783. doi:10.1056/NEJM197109302851405.

Flyamer, I. M., J. Gassler, M. Imakaev, H. B. Brandão, S. V. Ulianov, N. Abdennur, S. V. Razin, L. A. Mirny, and K. Tachibana-Konwalski. 2017. Single-nucleus Hi-C reveals unique chromatin reorganization at oocyte-to-zygote transition. *Nature* 544(7648):110–114. doi:10.1038/nature21711.

Fogarty, N. M. E., A. McCarthy, K. E. Snijders, B. E. Powell, N. Kubikova, P. Blakeley, R. Lea, K. Elder, S. E. Wamaitha, D. Kim, V. Maciulyte, J. Kleinjung, J. S. Kim, D. Wells, L. Vallier, A. Bertero, J. Turner, and K. K. Niakan. 2017. Genome editing reveals a role for OCT4 in human embryogenesis. *Nature* 550(7674):67–73. doi:10.1038/nature24033.

Frankel, M. S., and A. R. Chapman. 2000. *Human inheritable genetic modifications: Assessing scientific, ethical, religious and policy issues.* Washington, DC: American Association for the Advancement of Science.

Gao, L., K. Wu, Z. Liu, X. Yao, S. Yuan, W. Tao, L. Yi, G. Yu, Z. Hou, D. Fan, Y. Tian, J. Liu, Z. J. Chen, and J. Liu. 2018. Chromatin accessibility landscape in human early embryos and its association with evolution. *Cell* 173(1):248–259.e15. doi: 10.1016/j.cell.2018.02.028.

Gaudelli, N. M., D. K. Lam, H. A. Rees, N. M. Solá-Esteves, L. A. Barrera, D. A. Born, A. Edwards, J. M. Gehrke, S.-J. Lee, A. J. Liquori, R. Murray, M. S. Packer, C. Rinaldi, I. M. Slaymaker, J. Yen, L. E. Young, and G. Ciaramella. 2020. Directed evolution of adenine base editors with increased activity and therapeutic application. *Nature Biotechnology* 38:892–900. doi:10.1038/s41587-020-0491-6.

GEC (German Ethics Council). 2019. *Intervening in the human genome*. Berlin, Germany.

Genetic Alliance UK. 2016. *Genome Editing Technologies: The Patient Perspective*. London, U.K.

George, E. 2001. Beta-thalassemia major in Malaysia, an on-going public health problem. *Medical Journal of Malaysia* 60:397-400.

Golombok, S. 2017. Parenting in new family forms. Special issue edited by M. van IJzendoorn and M. Bakermans-Kranenburg. *Current Opinion in Psychology* 15:76–80. doi:10.1016/j.copsyc.2017.02.004.

Golombok, S. 2019. Parenting and contemporary reproductive technologies. In *Handbook of Parenting: Volume 3; Being and becoming a parent*, 3rd edition, edited by M. Bornstein. New York: Routledge.

Gorman, G. S., R. McFarland, J. Stewart, C. Feeney, and D. M. Turnbull. 2018. Mitochondrial donation: From test tube to clinic. *The Lancet* 392:1191–1192. doi:10.1016/S0140-6736(18)31868-3.

Graham, A., M. Powell, N. Taylor, D. Anderson, and R. Fitzgerald. 2013. *Ethical research involving children*. Florence, Italy: United Nations Children's Fund (UNICEF) Office of Research-Innocenti.

Greely, H. T. 2018. *The end of sex and the future of human reproduction*. Boston, MA: Harvard University Press.

Greenfield, A., P. Braude, F. Flinter, R. Lovell-Badge, C. Ogilvie, and A. C. F. Perry. 2017. Assisted reproductive technologies to prevent human mitochondrial disease transmission. *Nature Biotechnology* 35: 1059–1068

Gregoire, J., J. Georgas, D. H. Saklofske, F. Van de Vijver, C. Wierzbicki, L. G. Weiss, and J. Zhu. 2008. Cultural issues in the clinical use of the WISC-IV. In *WISC-IV clinical assessment and intervention*, edited by A. Prifitera, D. H. Saklofske, and L. G. Weiss. Amsterdam, Netherlands: Elsevier Academic Press. Pp. 517–544.

Griesinger, G., N. Bündgen, D. Salmen, E. Schwinger, G. Gillessen-Kaesbach, and K. Diedrich. 2009. Polar body biopsy in the diagnosis of monogenic diseases: The birth of three healthy children. *Deutsches Arzteblatt International* 106(33):533–538. doi:10.3238/arztebl.2009.0533.

Grünewald, J., R. Zhou, S. P. Garcia, S. Iyer, C. A. Lareau, M. J. Aryee and J. K. Joung. 2019. Transcriptome-wide off-target RNA editing induced by CRISPR-guided DNA base editors. *Nature* 569:433–437. doi:10.1038/s41586-019-1161-z.

Gu, B., E. Posfai, and J. Rossant. 2018. Efficient generation of targeted large insertions by microinjection into two-cell-stage mouse embryos. *Nature Biotechnology* 36:632–637. doi:10.1038/nbt.4166.

Gu, B., E. Posfai, M. Gertsenstein, and J. Rossant. 2020. Efficient generation of large-fragment knock-in mouse models using 2-cell (2C)-homologous recombination (HR)-CRISPR. *Current Protocols* 10(1):e67. doi:10.1002/cpmo.67.

GUaRDIAN Consortium, S. Sivasubbu, and V. Scaria. 2019. Genomics of rare genetic diseases: Experiences from India. *Human Genomics* 14(1):52. doi.org:10.1186/s40246-019-0215-5.

Guilinger, J. P., V. Pattanayak, D. Reyon, S. Q. Tsai, J. D. Sander, J. K. Joung, and D. R. Liu. 2014. Broad specificity profiling of TALENs results in engineered nucleases with improved DNA-cleavage specificity. *Nature Methods* 11(4):429–435. doi:10.1038/nmeth.2845.

Guo, H., P. Zhu, L. Yan, R. Li, B. Hu, Y. Lian, J. Yan, X. Ren, S. Lin, J. Li, X. Jin, X. Shi, P. Liu, X. Wang, W. Wang, Y. Wei, X. Li, F. Guo, X. Wu, X. Fan, J. Yong, L. Wen, S. X. Xie, F. Tang, and J. Qiao. 2014. The DNA methylation landscape of human early embryos. *Nature* 511:606–610. doi:10.1038/nature13544.

Handyside, A. H., E. H. Kontogianni, K. Hardy, and R. M. Winston. 1990. Pregnancies from biopsied human preimplantation embryos sexed by Y-specific DNA amplification. *Nature* 344(6268):768–770. doi:10.1038/344768a0.

Harbuz, R., R. Zouari, V. Pierre, M. Ben Khelifa, M. Kharouf, C. Coutton, G. Merdassi, F. Abada, J. Escoffier, Y. Nikas, F. Vialard, I. Koscinski, C. Triki, N. Sermondade, T. Schweitzer, A. Zhioua, F. Zhioua, H. Latrous, L. Halouani, M. Ouafi, M. Makni, P. S. Jouk, B. Sèle, S. Hennebicq, V. Satre, S. Viville, C. Arnoult, J. Lunardi, and P. F. Ray. 2011. A recurrent deletion of DPY19L2 causes infertility in man by blocking sperm head elongation and acrosome formation. *American Journal of Human Genetics* 88(3):351–361. doi:10.1016/j.ajhg.2011.02.007.

Hardy, K., A. H. Handyside, and R. M. Winston. 1989. The human blastocyst: Cell number, death, and allocation during late preimplantation development in vitro. *Development* 107(3):597–604.

Harper, J. C., K. Aittomaki, P. Borry, M. C. Cornel, G. de Wert, W. Dondorp, J. Geraedts, L. Gianaroli, K. Ketterson, I. Liebaers, K. Lundin, H. Mertes, M. Morris, G. Pennings, K. Sermon, C. Spits, S. Soini, A. P. A. van Montfoort, A. Veiga, J. R. Vermeesch, S. Viville, and M. Macek Jr., on behalf of the European Society of Human Reproduction and Embryology, and European Society of Human Genetics. 2018. Recent developments in genetics and medically assisted reproduction: From research to clinical applications. *European Journal of Human Genetics* 26(1):12–33. doi:10.1038/s41431-017-0016-z.

Hayashi, K., H. Ohta, K. Kurimoto, S. Aramaki, and M. Saitou. 2011. Reconstitution of the mouse germ cell specification pathway in culture by pluripotent stem cells. *Cell* 146(4):519–532. doi:10.1016/j.cell.2011.06.052.

Hayashi, K., S. Ogushi, K. Kurimoto, S. Shimamoto, H. Ohta, and M. Saitou. 2012. Offspring from oocytes derived from in vitro primordial germ cell-like cells in mice. *Science* 338(6109):971–975. doi:10.1126/science.1226889.

Heijligers, M., A. van Montfoort, M. Meijer-Hoogeveen, F. Broekmans, K. Bouman, I. Homminga, J. Dreesen, A. Paulussen, J. Engelen, E. Coonen, V. van der Schoot, M. van Deursen-Luijten, N. Muntjewerff, A. Peeters, R. van Golde, M. van der Hoeven, Y. Arens, and C. de Die-Smulders. 2018. Perinatal follow-up of children born after preimplantation genetic diagnosis between 1995 and 2014. *Journal of Assisted Reproduction and Genetics* 35(11):1995–2002. doi:10.1007/s10815-018-1286-2.

Hendriks, S., K. Peeraer, H. Bos, S. Repping, and E. A. F. Dancet. 2017. The importance of genetic parenthood for infertile men and women. *Human Reproduction* 32(10):2076–2087. doi:10.1093/humrep/dex256.

Henry, M. P., J. R. Hawkins, J. Boyle, and J. M. Bridger. 2019. The genomic health of human pluripotent stem cells: Genomic instability and the consequences on nuclear organization. *Frontiers in Genetics* 9:623. doi:10.3389/fgene.2018.00623.

Hermann, B. P., K. Cheng, A. Singh, L. Roa-De La Cruz, K. N. Mutoji, I. C. Chen, H. Gildersleeve, J. D. Lehle, M. Mayo, B. Westernströer, N. C. Law, M. J. Oatley, E. K. Velte, B. A. Niedenberger, D. Fritze, S. Silber, C. B. Geyer, J. M. Oatley, and J. R. McCarrey. 2018. The mammalian spermatogenesis single-cell transcriptome, from spermatogonial stem cells to spermatids. *Cell Reports* 25(6):1650–1667.e8. doi:10.1016/j.celrep.2018.10.026.

Heyer, W. D., K. T. Ehmsen, and J. Liu. 2010. Regulation of homologous recombination in eukaryotes. *Annual Review of Genetics* 44:113–139. doi: 10.1146/annurev-genet-0517 10-150955.

HFEA (Human Fertilisation and Embryology Authority). 2011. *Scientific review of the safety and efficacy of methods to avoid mitochondrial disease through assisted conception.* London, U.K. http://www.hfea.gov.uk/docs/2011-04- 18_Mitochondria_review_-_final_report.PDF.

HFEA. 2013. *Mitochondria replacement consultation: Advice to government.* London, U.K. https://www.hfea.gov.uk/media/2618/mitochondria_replacement_consultation_-_advice_for_government.pdf.

HFEA. 2014. *Third scientific review of the safety and efficacy of methods to avoid mitochondrial disease through assisted conception: 2014 update.* London, U.K. https://www.hfea.gov.uk/media/2614/third_mitochondrial_replacement_scientific_review.pdf.

HFEA. 2016. *Scientific review of the safety and efficacy of methods to avoid mitochondrial disease through assisted conception: 2016 update.* London, U.K. https://www.hfea.gov.uk/media/2611/fourth_scientific_review_mitochondria_2016.pdf.

HFEA. 2018. *Fertility treatment 2014-2016: Trends and figures.* London, U.K. https://www.hfea.gov.uk/media/2563/hfea-fertility-trends-and-figures-2017-v2.pdf.

Hikabe, O., N. Hamazaki, G. Nagamatsu, Y. Obata, Y. Hirao, N. Hamada, S. Shimamoto, T. Imamura, K. Nakashima, M. Saitou, and K. Hayashi. 2016. Reconstitution in vitro of the entire cycle of the mouse female germ line. *Nature* 539:299–303. doi:10.1038/nature20104.

Hinxton Group. 2015. *Statement on genome editing technologies and human germline genetic modification.* Baltimore, MD: The Hinxton Group. http://www.hinxtongroup.org/hinxton2015_statement.pdf.

Homfray, T., and P. A. Farndon. 2015. Chapter 7. Fetal anomalies: The geneticist's approach. In *Twining's textbook of fetal abnormalities*, 3rd edition, edited by A. M. Coady and S. Bower. Pp. 139–160.

Hou Y., W. Fan, L. Yan, R. Li, Y. Lian, J. Huang, J. Li, L. Xu, F. Tang, X. S. Xie, and J. Qiao. 2013. Genome analyses of single human oocytes. *Cell* 155(7):1492–506. doi:10.1016/j.cell.2013.11.040.

Hoy, S. 2019. Onasemnogene abeparvovec: First global approval. *Drugs* 79:1255–1262. doi: 10.1007/s40265-019-01162-5.

Hsu, P. D., E. S. Lander, and F. Zhang. 2014. Development and applications of CRISPR-Cas9 for genome engineering. *Cell* 157(6):1262–1278. doi:10.1016/j.cell.2014.05.010.

Huang, T. P., K. T. Zhao, S. M. Miller, N. M. Gaudelli, B. L. Oakes, C. Fellmann, D. F. Savage, and D. R. Liu. 2019. Circularly permuted and PAM-modified Cas9 variants broaden the targeting scope of base editors. *Nature Biotechnology* 37(6):626–631. doi:10.1038/s41587-019-0134-y.

Hurlbut, J. B., S. Jasanoff, K. Saha, A. Ahmed, A. Appiah, E. Bartholet, F. Baylis, G.Bennett, G. Church, I. G. Cohen, G. Daley, K. Finneran, W. Hurlbut, R. Jaenisch, L. Lwoff, J. P. Kimes, P. Mills, J. Moses, B. S. Park, E. Parens, R. Salzman, A. Saxena, H. Simmet, T. Simoncelli, O.C. Snead, K. Sunder Rajan, R. Truog, P. Williams, and C. Woopen. 2018. Building capacity for a global genome editing observatory: Conceptual challenges. *Trends in Biotechnology* 36(7):639–641. doi:10.1016/j.tibtech.2018.04.009.

Hustedt, N., and D. Durocher. 2017. The control of DNA repair by the cell cycle. *Nature Cell Biology* 19:1–9. doi: 10.1038/ncb3452.

Hyslop, L. A., P. Blakeley, L. Craven, J. Richardson, N. M. Fogarty, E. Fragouli, M. Lamb, S. E. Wamaitha, N. Prathalingam, Q. Zhang, H. O'Keefe, Y. Takeda, L. Arizzi, S. Alfarawati, H. A. Tuppen, L. Irving, D. Kalleas, M. Choudhary, D. Wells, A. P. Murdoch, D. M. Turnbull, K. K. Niakan, and M. Herbert. 2016. Towards clinical application of pronuclear transfer to prevent mitochondrial DNA disease. *Nature* 534(7607):383–386. doi: 10.1038/nature18303.

IFFS (International Federation of Fertility Societies). 2019. Global trends in reproductive policy and practice, 8th edition. *Global Reproductive Health* 4(1):e29 doi:10.1097/GRH.0000000000000029.

IHGSC (International Human Genome Sequencing Consortium). 2001. Initial sequencing and analysis of the human genome. *Nature* 409:860–921. doi: 10.1038/35057062.

IHGSC. 2004. Finishing the euchromatic sequence of the human genome. *Nature* 431:931–945 doi:10.1038/nature03001.

Isasi, R., E. Kleiderman, and B. M. Knoppers. 2016. Editing policy to fit the genome? *Science* 351(6271):337–339. doi:10.1126/science.aad6778.

Ishikura, Y., Y. Yabuta, H. Ohta, K. Hayashi, T. Nakamura, I. Okamoto, T. Yamamoto, K. Kurimoto, K. Shirane, H. Sasaki, and M. Saitou. 2016. In vitro derivation and propagation of spermatogonial stem cell activity from mouse pluripotent stem cells. *Cell Reports* 17(10):2789–2804. doi:10.1016/j.celrep.2016.11.026.

ISSCR (International Society for Stem Cell Research). 2015. *The ISSCR statement on human germline genome modification.* Skokie, IL. https://www.isscr.org/docs/default-source/policy-documents/isscr-statement-on-human-germline-genome-modification.pdf?sfvrsn=a34fb5bf_0.

ISSCR 2016. *Guidelines for stem cell research and clinical translation.* Skokie, IL. https://www.isscr.org/docs/default-source/all-isscr-guidelines/guidelines-2016/isscr-guidelines-for-stem-cell-research-and-clinical-translationd67119731dff6ddbb37cff0000940c19.pdf.

Jiang, W., S. Feng, S. Huang, W. Yu, G. Li, G. Yang, Y. Liu, Y. Zhang, L. Zhang, Y. Hou, J. Chen, J. Chen, and X. Huang. 2018. BE-PLUS: A new base editing tool with broadened editing window and enhanced fidelity. *Cell Research* 28(8):855–861. doi:10.1038/s41422-018-0052-4.

Joung, J. K., and J. D. Sander. 2013. TALENs: A widely applicable technology for targeted genome editing. *Nature Reviews Molecular Cell Biology* 14(1):49–55. doi:10.1038/nrm3486.

Kaiser, J. 2019. Update: House spending panel restores U.S. ban on gene-edited babies. *Science (news)*, June 4. doi:10.1126/science.aay1607.

Kang, J. G., J. S. Park, J. H. Ko, and Y. S. Kim. 2019. Regulation of gene expression by altered promoter methylation using a CRISPR/Cas9-mediated epigenetic editing system. *Scientific Reports* 9(1):11960. doi:10.1038/s41598-019-48130-3.

Karavani, E., O. Zuk, D. Zeevi, N. Barzilai, N. Stefanis, A. Hatzimanolis, N. Smyrnis, D. Avramopoulos, L. Kruglyak, G. Atzmon, M. Lam, T. Lencz, and S. Carmi. 2019. Screening human embryos for polygenic traits has limited utility. *Cell* 179(6):1424–1435. doi:10.1016/j.cell.2019.10.033.

Karvelis, T., G. Gasiunas, and V. Siksnys. 2017. Harnessing the natural diversity and in vitro evolution of Cas9 to expand the genome editing toolbox. *Current Opinion in Microbiology* 37:88–94. doi:10.1016/j.mib.2017.05.009.

Kasak, L., M. Punab, L. Nagirnaja, A. Grigorova, A. Minajeva, A. M. Lopes, A. M. Punab, K. I. Aston, F. Carvalho, E. Laasik, L. B. Smith, GEMINI Consortium, D. F. Conrad, and M. Laan. 2018. Bi-allelic recessive loss-of-function variants in FANCM cause non-obstructive azoospermia. *American Journal of Human Genetics* 103(2):200–212. doi.org/10.1016/j.ajhg.2018.07.005.

Kim, D., S. Bae, J. Park, E. Kim, S. Kim, H. R. Yu, J. Hwang, J. I. Kim, and J. S. Kim. 2015. Digenome-seq: Genome-wide profiling of CRISPR-Cas9 off-target effects in human cells. *Nature Methods* 12(3):237–243. doi:10.1038/nmeth.3284.

Kim, D., K. Luk, S. A. Wolfe, and J. S. Kim. 2019. Evaluating and enhancing target specificity of gene-editing nucleases and deaminases. *Annual Review of Biochemistry* 88:191–220. doi:10.1146/annurev-biochem-013118-111730.

Kim, Y. B., A. C. Komor, J. M. Levy, M. S. Packer, K. T. Zhao, and D. R. Liu. 2017. Increasing the genome-targeting scope and precision of base editing with engineered Cas9-cytidine deaminase fusions. *Nature Biotechnology* 35(4):371–376. doi:10.1038/nbt.3803.

Kleiderman, E., V. Ravitsky, and B. M. Knoppers. 2019. The "serious" factor in germline modification. *Journal of Medical Ethics* 45(8):508–513.

Kleinstiver, B. P., V. Pattanayak, M. S. Prew, S. Q. Tsai, N. T. Nguyen, Z. Zheng, and J. K. Joung. 2016. High-fidelity CRISPR-Cas9 nucleases with no detectable genome-wide off-target effects. *Nature* 529(7587):490–495. doi:10.1038/nature16526.

KNAW (Royal Netherlands Academy of Arts and Sciences). 2016. Genome editing: Position paper of the Royal Netherlands Academy of Arts and Sciences. Amsterdam, Netherlands. https://www.knaw.nl/en/news/publications/genome-editing.

Kosicki, M., K. Tomberg, and A. Bradley. 2018. Repair of double-strand breaks induced by CRISPR–Cas9 leads to large deletions and complex rearrangements. *Nature Biotechnology* 36:765–771. doi:10.1038/nbt.4192.

Krausz, C., and A. Riera-Escamilla. 2018. Genetics of male infertility. *Nature Reviews Urology* 15:369–384. doi:10.1038/s41585-018-0003-3.

Kubota, H., and Brinster, R. L. 2018. Spermatogonial stem cells. *Biology of Reproduction* 99(1):52–74. doi:10.1093/biolre/ioy077.

Kuiper, D., A. Bennema, S. la Bastide-van Gemert, J. Seggers, P. Schendelaar, S. Mastenbroek, A. Hoek, M. J. Heineman, T. J. Roseboom, J. H. Kok, and M. Hadders-Algra. 2018. Developmental outcome of nine-year-old children born after PGS: Follow-up of a randomized trial. *Human Reproduction* 33(1):147–155. doi:10.1093/humrep/dex337.

Kurt, I. C., R. Zhou, S. Iyer, S. P. Garcia, B. R. Miller, L. M. Langner, J. Grünewald, and J. K. Joung. 2020. CRISPR C-to-G base editors for inducing targeted DNA transversions in human cells. *Nature Biotechnology.* doi:10.1038/s41587-020-0609-x.

Lander, E. S., F. Baylis, F. Zhang, E. Charpentier, P. Berg, C. Bourgain, B. Friedrich, J. K. Joung, J. Li, D. Liu, L. Naldini, J. B. Nie, R. Qiu, B. Schoene-Seifert, F. Shao, S. Terry, W. Wei, and E. L. Winnacker. 2019. Adopt a moratorium on heritable genome editing. *Nature* 567(7747):165–168. doi:10.1038/d41586-019-00726-5.

Lanphier, E., F. Urnov, S. E. Haecker, M. Werner, and J. Smolenski. 2015. Don't edit the human germ line. *Nature* 519(7544):410–411. doi:10.1038/519410a.

Lea, R., and K. Niakan. 2019. Human germline genome editing. *Nature Cell Biology* 21(12):1479–1489. doi:10.1038/s41556-019-0424-0.

Leaver, M., and D. Wells. 2020. Non-invasive preimplantation genetic testing (niPGT): The next revolution in reproductive genetics? *Human Reproduction Update* 26(1):16–42. doi:10.1093/humupd/dmz033.

Lee, H. K., H. E. Smith, C. Liu, M. Willi, and L. Hennighausen. 2020. Cytosine base editor 4 but not adenine base editor generates off-target mutations in mouse embryos. *Communications Biology* 3:19. doi:10.1038/s42003-019-0745-3.

Leopoldina (Leopoldina, Acatech, DFG, and Academien Union). 2015. *The opportunities and limits of genome editing.* https://www.leopoldina.org/en/publications/detailview/publication/chancen-und-grenzen-des-genome-editing-2015/.

Li, F., Z. An, and Z. Zhang. 2019. The dynamic 3D genome in gametogenesis and early embryonic development. *Cells* 8(8):788. doi:10.3390/cells8080788.

Li, G., Y. Liu, Y. Zeng, J. Li, L. Wang, G. Yang, D. Chen, X. Shang, J. Chen, X. Huang, and J. Liu. 2017. Highly efficient and precise base editing in discarded human tripronuclear embryos. *Protein & Cell* 8(10):776–779. doi:10.1007/s13238-017-0458-7.

Li, H., Y. Yang, W. Hong, M. Huang, M. Wu, and X. Zhao. 2020. Applications of genome editing technology in the targeted therapy of human diseases: Mechanisms, advances and prospects. *Signal Transduction and Targeted Therapy* 5:1. doi:10.1038/s41392-019-0089-y.

Li, L., F. Guo, Y. Gao, Y. Ren, P. Yuan, L. Yan, R. Li, Y. Lian, J. Li, B. Hu, J. Gao, L. Wen, F. Tang, and J. Qiao. 2018. Single-cell multi-omics sequencing of human early embryos. *Nature Cell Biology* 20:847–858. doi:10.1038/s41556-018-0123-2.

Liang, D., N. M. Gutierrez, T. Chen, Y. Lee, , S. Park, H. Ma, A. Koski, R. Ahmed, H. Darby, Y. Li, C. Van Dyken, A. Mikhalchenko, T. Gonmanee, T. Hayama, H. Zhao, K. Wu, J. Zhang, Z. Hou, J. Park, C. Kim, J. Gong, Y. Yuan, Y. Gu, Y. Shen, S. B. Olson, H. Yang, D. Battaglia, T. O'Leary, S. A. Krieg, D. M. Lee, D. H. Wu, P. B. Duell, S. Kual, J. Kim, S. B. Heitner, E. Kang, Z. Chen, P. Amato, and S. Mitalipov. 2020 Frequent gene conversion in human embryos induced by double strand breaks. *bioRxiv* 2020.06.19.162214. doi:10.1101/2020.06.19.162214.

Liang, P., Y. Xu, X. Zhang, C. Ding, R. Huang, Z. Zhang, J. Lv, X. Xie, Y. Chen, Y. Li, Y. Sun, Y. Bai, Z. Songyang, W. Ma, C. Zhou, and J. Huang. 2015. CRISPR/Cas9-mediated gene editing in human tripronuclear zygotes. *Protein and Cell* 6(5):363–372. doi:10.1007/s13238-015-0153-5.

Liu, L., L. Leng, C. Liu, C. Lu, Y. Yuan, L. Wu, F. Gong, S. Zhang, X. Wei, M. Wang, L. Zhao, L. Hu, J. Wang, H. Yang, S. Zhu, F. Chen, G. Lu, Z. Shang, and G. Lin. 2019a. An integrated chromatin accessibility and transcriptome landscape of human pre-implantation embryos. *Nature Communications* 10:364. doi:10.1038/s41467-018-08244-0.

Liu, M., S. Rehman, X. Tang, K. Gu, Q. Fan, D. Chen, and W. Ma. 2019b. Methodologies for improving HDR efficiency. *Frontiers in Genetics* 9:691. doi:10.3389/fgene.2018.00691.

Liu, Y., X. Li, S. He, S. Huang, C. Li, Y. Chen, Z. Liu, X. Huang and X. Wang. 2020. Efficient generation of mouse models with the prime editing system. *Cell Discovery* 6:27. doi:10.1038/s41421-020-0165-z.

Lochmüller, H., J. Torrent i Farnell, Y. Le Cam, A. H. Jonker, L. P. L. Lau, G. Baynam, P. Kaufmann, H. J. S. Dawkins, P. Lasko, C. P. Austin, and K. M. Boycott, on behalf of the IRDiRC Consortium Assembly. 2017. The International Rare Diseases Research Consortium: Policies and guidelines to maximize impact. *European Journal of Human Genetics* 25:1293–1302. doi.org/10.1038/s41431-017-0008-z.

Ma, F., Y. Yang, X. Li, F. Zhou, C. Gao, M. Li, and L. Gao. 2013. The association of sport performance with ACE and ACTN3 genetic polymorphisms: A systematic review and meta-analysis. *PLoS One* 8(1):e54685. doi:10.1371/journal.pone.0054685.

Ma, H., N. Marti-Gutierrez, S. W. Park, J. Wu, T. Hayama, H. Darby, C. Van Dyken, Y. Li, A. Koski, D. Liang, K. Suzuki, Y. Gu, J. Gong, X. Xu, R. Ahmed, Y. Lee, E. Kang, D. Ji, A. R. Park, D. Kim, S.-T. Kim, S. B. Heitner, D. Battaglia, S. A. Krieg, D. M. Lee, D. H. Wu, D. P. Wolf, P. Amato, S. Kaul, J. C. Izpisua Belmonte, J.-S. Kim, and S. Mitalipov. 2018. Ma et al. reply. *Nature* 560:E10–E23. doi:10.1038/s41586-018-0381-y.

Ma, H., N. Marti-Gutierrez, S. W. Park, J. Wu, Y. Lee, K. Suzuki, A. Koski, D. Ji, T. Hayama, R. Ahmed, H. Darby, C. Van Dyken, Y. Li, E. Kang, A. R. Park, D. Kim, S. T. Kim, J. Gong, Y. Gu, X. Xu, D. Battaglia, S. A. Krieg, D. M. Lee, D. H. Wu, D. P. Wolf, S. B. Heitner, J. C. I. Belmonte, P. Amato, J. S. Kim, S. Kaul, and S. Mitalipov. 2017. Correction of a pathogenic gene mutation in human embryos. *Nature* 548(7668):413–419. doi:10.1038/nature23305.

Maor-Sagie, E., Y. Cinnamon, B. Yaacov, A. Shaag, H. Goldsmidt, S. Zenvirt, N. Laufer, C. Richler, and A. Frumkin. 2015. Deleterious mutation in SYCE1 is associated with non-obstructive azoospermia. *Journal of Assisted Reproduction and Genetics* 32(6):887–891. doi.org/10.1007/s10815-015-0445-y.

McCann, J. L., D. J. Salamango, E. K. Law, W. L. Brown, , and R. S. Harris. 2020. MagnEdit-interacting factors that recruit DNA-editing enzymes to single base targets. *Life Science Alliance* 3(4):e201900606. doi:10.26508/lsa.201900606.

Mianné, J., G. F. Codner, A. Caulder, R. Fell, M. Hutchison, R. King, M. E. Stewart, S. Wells, and L. Teboul. 2017. Analysing the outcome of CRISPR-aided genome editing in embryos: Screening, genotyping and quality control. *Methods* 15:121–122:68-76. doi:10.1016/j.ymeth.2017.03.016.

Migeon, B.R. 2020. X-linked diseases: Susceptible females. *Genetics in Medicine* 22:1156–1174. doi:10.1038/s41436-020-0779-4.

Mok, B.Y., M. H. de Moraes, J. Zeng, D. E. Bosch, A. V. Kotrys, A. Raguram, F. Hsu, M. C. Radey, S. Brook Peterson, V. K. Mootha, J. D. Mougous and D. R. Liu. 2020. A bacterial cytidine deaminase toxin enables CRISPR-free mitochondrial base editing. *Nature* 583:631–637. doi:10.1038/s41586-020-2477-4.

Moris, N., K. Anlas, S. C. van den Brink, A. Alemany, J. Schröder, S. Ghimire, T. Balayo, A. van Oudenaarden and A. M. Arias. 2020. An in vitro model of early anteroposterior organization during human development. *Nature* 582:410–415. doi:10.1038/s41586-020-2383-9.

Morohaku, K., R. Tanimoto, K. Sasaki, R. Kawahara-Miki, T. Kono, K. Hayashi, Y. Hirao, and Y. Obata. 2016. Complete in vitro generation of fertile oocytes from mouse primordial germ cells. *PNAS* 113(32):9021–9026. https://doi.org/10.1073/pnas.1603817113.

Nagamatsu, G., and K. Hiyashi. 2017. Stem cells, in vitro gametogenesis, and male fertility. *Reproduction* 154(6):F79–F91. doi: 10.1530/REP-17-0510.

NASEM (National Academies of Sciences, Engineering, and Medicine). 2015. *International summit on human gene editing: A global discussion.* Washington, DC: The National Academies Press.

NASEM. 2016. *Mitochondrial replacement techniques: Ethical, social, and policy considerations.* Washington, DC: The National Academies Press.

NASEM. 2017. *Human genome editing: Science, ethics, and governance.* Washington, DC: The National Academies Press.

NASEM. 2019a. *Framework for addressing ethical dimensions of emerging and innovative biomedical technologies: A synthesis of relevant National Academies reports.* Washington, DC: The National Academies Press. doi:10.17226/25491.

NASEM. 2019b. *Second international summit on human genome editing: Continuing the global discussion; Proceedings of a workshop–in brief.* Washington, DC: The National Academies Press.

Nuffield Council on Bioethics (NCB). 2012. *Novel techniques for the prevention of mitochondrial DNA disorders: An ethical review.* London, U.K. https://www.nuffieldbioethics.org/assets/pdfs/Novel_techniques_for_the_prevention_of_mitochondrial_DNA_disorders.pdf.

Nuffield Council on Bioethics. 2016. *Genome editing: An ethical review.* London, U.K.

Nuffield Council on Bioethics. 2018. *Genome editing and human reproduction: Social and ethical issues.* London, U.K.

Niakan, K. 2019. Mechanisms of lineage specification in human embryos. Presentation to the International Commission on the Clinical Use of Human Germline Genome Editing, November 14. https://www.nationalacademies.org/event/11-14-2019/international-commission-on-the-clinical-use-of-human-germline-genome-editing-commission-meeting-2. Accessed July 7, 2020.

Niakan, K. K., and K. Eggan. 2013. Analysis of human embryos from zygote to blastocyst reveals distinct gene expression patterns relative to the mouse. *Developmental Biology* 375(1):54–64. doi:10.1016/j.ydbio.2012.12.008.

Niu, Y., B. Shen, Y. Cui, Y. Chen, J. Wang, L. Wang, Y. Kang, X. Zhao, W. Si, W. Li, A. P. Xiang, J. Zhou, X. Guo, Y. Bi, C. Si, B. Hu, G. Dong, H. Wang, Z. Zhou, T. Li, T. Tan, X. Pu, F. Wang, S. Ji, Q. Zhou, X. Huang, W. Ji, and J. Sha. 2014. Generation of gene-modified cynomolgus monkey via Cas9/RNA-mediated gene targeting in one-cell embryos. *Cell* 156(4):836–843. doi:10.1016/j.cell.2014.01.027.

Normile, D. 2019. Chinese scientist who produced genetically altered babies sentenced to three years in jail. *Science (news)*, December 30. doi:10.1126/science.aba7347.

Nsota Mbango, J. F., C. Coutton, C. Arnoult, P. F. Ray, and A. Touré. 2019. Genetic causes of male infertility: Snapshot on morphological abnormalities of the sperm flagellum. *Basic and Clinical Andrology* 29:2. doi:10.1186/s12610-019-0083-9.

Okutman, O., J. Muller, Y. Baert, M. Serdarogullari, M. Gultomruk, A. Piton, C. Rombaut, M. Benkhalifa, M. Teletin, V. Skory, E. Bakircioglu, E. Goossens, M. Bahceci, and S. Viville. 2015. Exome sequencing reveals a nonsense mutation in *TEX15* causing spermatogenic failure in a Turkish family. *Human Molecular Genetics* 24(19):5581–5588. doi: 10.1093/hmg/ddv290.

Oprea, T. I., L. Jan, G. L. Johnson, B. L. Roth, A. Ma'ayan, S. Schürer, B. K. Shoichet, L. A. Sklar, and M. T. McManus. 2018. Far away from the lamppost. *PLOS Biology* 16(12):e3000067. doi:10.1371/journal.pbio.3000067.

Padden, C., and J. Humphries. 2020. Who goes first? Deaf people and CRISPR germline editing. *Perspectives in Biology and Medicine* 63(1):54–65. https://muse.jhu.edu/article/748050. doi:10.1353/pbm.2020.0004.

Paix, A., A. Folkmann, D. H. Goldman, H. Kulaga, M. J. Grzelak, D. Rasoloson, S. Paidemarry, R. Green, R. R. Reed, and G. Seydoux. 2017. Precision genome editing using synthesis-dependent repair of Cas9-induced DNA breaks. *Proceedings of the National Academy of Sciences of the United States of America* 114(50):E10745–E10754. doi:10.1073/pnas.1711979114.

Paradiñas, A. F., P. Holmans, A. J. Pocklington, V. Escott-Price, S. Ripke, N. Carrera, S. E. Legge, S. Bishop, D. Cameron, M. L. Hamshere, J. Han, L. Hubbard, A. Lynham, K. Mantripragada, E. Rees, J. H. MacCabe, S. A. McCarroll, B. T. Baune, G. Breen, E. M. Byrne, U. Dannlowski, T. C. Eley, C. Hayward, N. G. Martin, A. M. McIntosh, R. Plomin, D. J. Porteous, N. R. Wray, A. Caballero, D. H. Geschwind, L. M. Huckins, D. M. Ruderfer, E. Santiago, P. Sklar, E. A. Stahl, H. Won, E. Agerbo, T. D. Als, O. A. Andreassen, M. Bækvad-Hansen, P. B. Mortensen, C. B. Pedersen, A. D. Børglum, J. Bybjerg-Grauholm, S. Djurovic, N. Durmishi, M. G. Pedersen, V. Golimbet, J. Grove, D. M. Hougaard, M. Mattheisen, E. Molden, O. Mors, M. Nordentoft, M. Pejovic-Milovancevic, E. Sigurdsson, T. Silagadze, C. S. Hansen, K. Stefansson, H. Stefansson, S. Steinberg, S. Tosato, T. Werge, GERAD1 Consortium, CRESTAR Consortium, D. A. Collier, D. Rujescu, G. Kirov, M. J. Owen, M. C. O'Donovan, and J. T. R. Walters. 2018. Common schizophrenia alleles are enriched in mutation-intolerant genes and in regions under strong background selection. *Nature Genetics* 50(3):381–389. doi: 10.1038/s41588-018-0059-2.

Pickering, C., and J. Kiely. 2017. ACTN3: More than just a gene for speed. *Frontiers in Physiology* 8:1080. doi:10.3389/fphys.2017.01080.

Platt, O. S., D. J. Brambilla, W. F. Rosse, P. F. Milner, O. Castro, M. H. Steinberg, and P. P. Klug. 1994. Mortality in sickle cell disease: Life expectancy and risk factors for early death. *New England Journal of Medicine* 330:1639–1644. doi:10.1056/NEJM199406093302303.

Posey, J. E., A. H. O'Donnell-Luria, J. X. Chong, T. Harel, S. N. Jhangiani, Z. H. Coban Akdemir, S. Buyske, D. Pehlivan, C. Carvalho, S. Baxter, N. Sobreira, P. Liu, N. Wu, J. A. Rosenfeld, S. Kumar, D. Avramopoulos, J. J. White, K. F. Doheny, P. D. Witmer, C. Boehm, V. R. Sutton, D. M. Muzny, E. Boerwinkle, M. Günel, D. A. Nickerson, S. Mane, D. G. MacArthur, R. A. Gibbs, A. Hamosh, R. P. Lifton, T. C. Matise, H. L. Rehm, M. Gerstein, M. J. Bamshad, D. Valle, J. R. Lupski, and Centers for Mendelian Genomics. 2019. Insights into genetics, human biology and disease gleaned from family-based genomic studies. *Genetics in Medicine* 21(4):798–812. doi:10.1038/s41436-018-0408-7.

President's Commission. 1982. *Splicing life: A report on the social and ethical issues of genetic engineering with human beings.* Washington, DC: President's Commission for the Study of Ethical Problems in Medicine and Biomedical and Behavioral Research. https://bioethics.georgetown.edu/documents/pcemr/splicinglife.pdf.

Pulecio, J., N. Verma, E. Mejía-Ramírez, D. Huangfu, and A. Raya. 2017. CRISPR/Cas9-Based Engineering of the Epigenome. *Cell Stem Cell* 21(4):431–447. doi:10.1016/j.stem.2017.09.006.

Quinn, C. T., Z. R. Rogers, T. L. McCavit, and G. R. Buchanan. 2010. Improved survival of children and adolescents with sickle cell disease. *Blood* 115(17):3447–3452. doi: 10.1182/blood-2009-07-233700.

Rasmussen, K. L., A. Tybjærg-Hansen, B. G. Nordestgaard, and R. Frikke-Schmidt. 2018. Absolute 10-year risk of dementia by age, sex, and APOE genotype: A population-based cohort study. *Canadian Medical Association Journal* 90(35):E1033–E1041. doi:10.1503/cmaj.180066.

RCOG (Royal College of Obstetricians and Gynaecologists). 2016. Ovarian Hyperstimulation Syndrome. London, U.K. https://www.rcog.org.uk/globalassets/documents/patients/patient-information-leaflets/gynaecology/pi_ohss.pdf.

Reddy P., A. Ocampo, K. Suzuki, J. Luo, S. R. Bacman, S. L. Williams, A. Sugawara, D. Okamura, Y. Tsunekawa, J. Wu, D. Lam, X. Xiong, N. Montserrat, C. R. Esteban, G. H. Liu, I. Sancho-Martinez, D. Manau, S. Civico, F. Cardellach, M. Del Mar O'Callaghan, J. Campistol, H. Zhao, J. M. Campistol, C. T. Moraes, and J. C. Izpisua Belmonte. 2015. Selective elimination of mitochondrial mutations in the germline by genome editing. *Cell* 161(3):459–469. doi:10.1016/j.cell.2015.03.051.

Rees, H. A., and D. R. Liu. 2018. Base editing: Precision chemistry on the genome and transcriptome of living cells. *Nature Reviews Genetics* 19(12):770–788. doi:10.1038/s41576-018-0059-1.

Richter, M. F., K. T. Zhao, E. Eton, A. Lapinaite, G. A. Newby, B. W. Thuronyi, C. Wilson, L. W. Koblan, J. Zeng, D. E. Bauer, J. A. Doudna and D. R. Liu. 2020. Phage-assisted evolution of an adenine base editor with improved Cas domain compatibility and activity. *Nature Biotechnology* 38:883–891. doi:10.1038/s4157-020-0414-6.

Riordan, J. R., J. M. Rommens, B. Kerem, N. Alon, R. Rozmahel, Z. Grzelczak, J. Zielenski, S. Lok, N. Plavsic, J. L. Chou, M. L. Drumm, M. C. Iannuzzi, F. S. Collins, and L.-C. Tsui. 1989. Identification of the cystic fibrosis gene: Cloning and characterization of complementary DNA. *Science* 245(4922):1066–1073. doi:10.1126/science.2475911.

Rivron, N., M. Pera, J. Rossant, A. Martinez Arias, M. Zernicka-Goetz, J. Fu, S. van den Brink, A. Bredenoord, W. Dondorp, G. de Wert, I. Hyun, M. Munsie, and R. Isasi. 2018. Debate ethics of embryo models from stem cells. *Nature* 564(7735):183–185. doi:10.1038/d41586-018-07663-9.

Rockoff, J. D. 2019. New gene therapy priced at $1.8 million in Europe. *Wall Street Journal*, June 14. https://www.wsj.com/articles/new-gene-therapy-priced-at-1-8-million-in-europe-11560529116.

Romdhane, L., N. Mezzi, Y. Hamdi, G. El-Kamah, A. Barakat, and S. Abdelhak. 2019. Consanguinity and inbreeding in health and disease in North African populations. *Annual Review of Genomics and Human Genetics* 20:155–179.

Rossant, J., and P. P. L. Tam. 2017. New insights into early human development: Lessons for stem cell derivation and differentiation. *Cell Stem Cell* 20:18-28. doi: 10.1016/j.stem.2016.12.004.

Rouet, P., F. Smih, and M. Jasin. 1994. Introduction of double-strand breaks into the genome of mouse cells by expression of a rare-cutting endonuclease. *Molecular and Cellular Biology* 14(12):8096–8106. doi:10.1128/mcb.14.12.8096.

Rulli, T. 2014. Preferring a genetically-related child. *Journal of Moral Philosophy* 1-30. doi:10.1163/17455243-4681062.

Sakuma, T., S. Nakade, Y. Sakane, K.-I. T. Suzuki and T. Yamamoto. 2016. MMEJ-assisted gene knock-in using TALENs and CRISPR-Cas9 with the PITCh systems. *Nature Protocols* 11:118–133. doi:10.1038/nprot.2015.140

Sander, J. D., and J. K. Joung. 2014. CRISPR-Cas systems for editing, regulating and targeting genomes. *Nature Biotechnology* 32(4):347–355. doi:10.1038/nbt.2842.

Sasaki, K., S. Yokobayashi, T. Nakamura, I. Okamoto, Y. Yabuta, K. Kurimoto, H. Ohta, Y. Moritoki, C. Iwatani, H. Tsuchiya, S. Nakamura, K. Sekiguchi, T. Sakuma, T. Yamamoto, T. Mor, K. Woltjen, M. Nakagawa, T. Yamamoto, K. Takahashi, S. Yamanaka, and M. Saitou. 2015. Robust in vitro induction of human germ cell fate from pluripotent stem cells. *Cell Stem Cell* 7(2):178–194. doi:10.1016/j.stem.2015.06.014.

Sasani, T. A., B. S. Pedersen, Z. Gao, L. Baird, M. Przeworski, L. B. Jorde, and A. R. Quinlan. 2019. Large, three-generation human families reveal post-zygotic mosaicism and variability in germline mutation accumulation. *eLife* 8:e46922. doi: 10.7554/eLife.46922.

Schenk, M., A. Groselj-Strele, K. Eberhard, E. Feldmeier, D. Kastelic, S. Cerk, and G. Weiss. 2018. Impact of polar body biopsy on embryo morphokinetics-back to the roots in preimplantation genetic testing? *Journal of Assisted Reproduction and Genetics* 35(8):1521–1528. doi:10.1007/s10815-018-1207-4.

Schultz, N., F. K. Hamra, and D. L. Garbers. 2003. A multitude of genes expressed solely in meiotic or postmeiotic spermatogenic cells offers a myriad of contraceptive targets. *Proceedings of the National Academy of Sciences* 100(21):12201–12206. doi:10.1073/PNAS.1635054100.

Segers, S., G. Pennings, and H. Mertes. 2019. Getting what you desire: The normative significance of genetic relatedness in parent-child relationships. *Medicine, Health Care and Philosophy* 22:487–495. doi:10.1007/s11019-019-09889-4.

Simunovic, M., and A. H. Brivanlou. 2017. Embryoids, organoids, and gastruloids: New approaches to understanding embryogenesis. *Development* 144(6):976–985. doi:10.1242/dev.143529.

Slaymaker, I. M., L. Gao, B. Zetsche, D. A. Scott, W. X. Yan, and F. Zhang. 2016. Rationally engineered Cas9 nucleases with improved specificity. *Science* 351(6268):84–88. doi:10.1126/science.aad5227.

Smith, Z. D., M. M. Chan, K. C. Humm, R. Karnik, S. Mekhoubad, A. Regev, K. Eggan, and A. Meissner. 2014. DNA methylation dynamics of the human preimplantation embryo. *Nature* 511(7511):611–615. doi:10.1038/nature13581.

SRCD (Society for Research in Child Development). 2007. Ethical standards for research with children. https://www.srcd.org/about-us/ethical-standards-research-children. Accessed November 16, 2020.

Stadtmauer, E. A., J. A. Fraietta, M. M. Davis, A. D. Cohen, K. Weber, E. Lancaster, P. A. Mangan, I. Kulikovskaya, M. Gupta, F. Chen, L. Tian, V. E. Gonzalez, J. Xu, I. Y. Jung, J. J. Melenhorst, G. Plesa, J. Shea, T. Matlawski, A. Cervini, A. L. Gaymon, S. Desjardins, A. Lamontagne, J. Salas-Mckee, A. Fesnak, D. L. Siegel, B. L. Levine, J. K. Jadlowsky, R. M. Young, A. Chew, W. T. Hwang, E. O. Hexner, B. M. Carreno, C. L. Nobles, F. D. Bushman, K. R. Parker, Y. Qi, A. T. Satpathy, H. Y. Chang, Y. Zhao, S. F. Lacey, and C. H. June. 2020. CRISPR-engineered T cells in patients with refractory cancer. *Science* 367(6481):eaba7365. doi:10.1126/science.aba7365.

Steffann, J., P. Jouannet, J. P. Bonnefont, H. Chneiweiss, and N. Frydman. 2018. Could failure in preimplantation genetic diagnosis justify editing the human embryo genome? *Cell Stem Cell* 22(4):481–482. doi:10.1016/j.stem.2018.01.004.

Stock, G., and J. Campbell. 2000. *Engineering the human germline*. Oxford, U.K.: Oxford University Press.

Strom, C. M., B. Crossley, A. Buller-Buerkle, M. Jarvis, F. Quan, M. Peng, K. Muralidharan, V. Pratt, J. B. Redman, and W. Sun. 2011. Cystic fibrosis testing eight years on: Lessons learned from carrier screening and sequencing analysis. *Genetics in Medicine* 13:166–172. doi:10.1097/GIM.0b013e3181fa24c4.

Sürün, D., A. Schneider, J. Mircetic, K. Neumann, F. Lansing, M. Paszkowski-Rogacz, V. Hänchen, M.A. Lee-Kirsch, and F. Buchholz. 2020. Efficient generation and correction of mutations in human iPS cells utilizing mRNAs of CRISPR base editors and prime editors. *Genes* 11:511. doi:10.3390/genes11050511.

Tang, L., Y. Zeng, X. Zhou, H. Du, C. Li, J. Liu, and P. Zhang. 2018. Highly efficient ssODN-mediated homology-directed repair of DSBs generated by CRISPR/Cas9 in human 3PN zygotes. *Molecular Reproduction and Development* 85(6):461–463. doi: 10.1002/CD4.22983.

Tebas, P., D. Stein, W. W. Tang, I. Frank, S. Q. Wang, G. Lee, S. K. Spratt, R. T. Surosky, M. A. Giedlin, G. Nichol, M. C. Holmes, P. D. Gregory, D. G. Ando, M. Kalos, R. G. Collman, G. Binder-Scholl, G. Plesa, W. T. Hwang, B. L. Levine, and C. H. June. 2014. Gene editing CD4CCR5 in autolCD4us CD4 T cells of persons infected with HIV. *New England Journal of Medicine* 370(10):901–910. doi:10.1056/NEJMoa1300662.

Tenenbaum-Rakover, Y., A. Weinberg-Shukron, P. Renbaum, O. Lobel, H. Eideh, S. Gulsuner, D. Dahary, A. Abu-Rayyan, M. Kanaan, E. Levy-Lahad, D. Bercovich, and D. Zangen. 2015. Minichromosome maintenance complex component 8 (MCM8) gene mutations result in primary gonadal failure. *Journal of Medical Genetics* 52:391–399.

Timpson, N. J., C. M. T. Greenwood, N. Soranzo, D. J. Lawson, and J. B. Richards. 2018. Genetic architecture: The shape of the genetic contribution to human traits and disease. *Nature Reviews Genetics* 19(2):110–124. doi:10.1038/nrg.2017.101.

Tsai, S. Q., and J. K. Joung. 2016. Defining and improving the genome-wide specificities of CRISPR-Cas9 nuclease. *Nature Reviews Genetics* 17(5):300–312. doi:10.1038/nrg.2016.28.

UKDH (United Kingdom Department of Health). 2000. Stem cell research: Medical progress with responsibility. *Cloning* 2(2):91–96. doi:10.1089/152045500436113.

UNESCO (United Nations Educational, Scientific, and Cultural Organization). 2015. *Report of the International Bioethics Committee on updating its reflection on the human genome and human rights.* Paris, France. http://www.coe.int/en/web/bioethics/-/gene-editing.

Viotti, M., A. R. Victor, D. K. Griffin, J. S. Groob, A. J. Brake, C. G. Zouves, and F. L. Barnes. 2019. Estimating demand for germline genome editing: An in vitro fertilization clinic perspective. *The CRISPR Journal* 2(5):304–315. doi: 10.1089/crispr.2019.0044.

Walker, F. O. 2007. Huntington's disease. *The Lancet* 369(9557):218–228. doi:10.1016/S0140-6736(07)60111-1.

Wang, L., and J. Li. 2019. "Artificial spermatid"–mediated genome editing dagger. *Biology of Reproduction* 101:538–548. doi:10.1093/biolre/ioz087.

Wang, Y., Q. Liu, F. Tang, L. Yan, and J. Qiao. 2019. Epigenetic regulation and risk factors during the development of human gametes and early embryos. *Annual Review of Genomics and Human Genetics* 20:21–40. doi:10.1146/annurev-genom-083118-015143.

Warmflash, A. 2017. Synthetic embryos: Windows into mammalian development. *Cell Stem Cell* 20(5):581–582. doi:10.1016/j.stem.2017.04.001.

Wensink, P. C., D. J. Finnegan, J. E. Donelson, and D. S. Hogness. 1974. A system for mapping DNA sequences in the chromosomes of Drosophila melanogaster. *Cell* 3(4):315–325. doi:10.1016/0092-8674(74)90045-2.

Wertz, D. C., and B. M. Knoppers. 2002. Serious genetic disorders: Can or should they be defined? *American Journal of Medical Genetics* 108(1):29–35. doi:10.1002/ajmg.10212.

WHO (World Health Organization). 2019a. *Genes and human diseases.* Geneva, Switzerland. https://www.who.int/genomics/public/geneticdiseases/en/index2.html.

WHO. 2019b. WHO launches global registry on human genome editing. News release, August 29. Geneva, Switzerland. https://www.who.int/news-room/detail/29-08-2019-who-launches-global-registry-on-human-genome-editing.

WHO. 2020. *Human Genome Editing: A DRAFT Framework for Governance,* July 3, 2020. Available at: https://www.who.int/docs/default-source/ethics/governance-framework-forhuman-genome-editing-2ndonlineconsult.pdf?ua=1.

Wienert, B., S. K. Wyman, C. D. Richardson, C. D. Yeh, P. Akcakaya, M. J. Porritt, M. Morlock, J. T. Vu, K. R. Kazane, H. L. Watry, L. M. Judge, B. R. Conklin, M. Maresca, and J. E. Corn. 2019. Unbiased detection of CRISPR off-targets in vivo using DISCOVER-Seq. *Science* 364(6437):286–289. doi:10.1126/science.aav9023.

WMA (World Medical Association). 2013. World Medical Association Declaration of Helsinki: Ethical principles for medical research involving human subjects. *Journal of the American Medical Association* 310(20):2191–2194. doi:10.1001/jama.2013.281053.

Wu, Y., H. Zhou, X. Fan, Y. Zhang, M. Zhang, Y. Wang, Z. Xie, M. Bai, Q. Yin, D. Liang, W. Tang, J. Liao, C. Zhou, W. Liu, P. Zhu, H. Guo, H. Pan, C. Wu, H. Shi, L. Wu, F. Tang, and J. Li. 2015. Correction of a genetic disease by CRISPR-Cas9-mediated gene editing in mouse spermatogonial stem cells. *Cell Research* 25:67–79. doi:10.1038/cr.2014.160.

Xu, Q., and W. Xie. 2018. Epigenome in early mammalian development: Inheritance, re-programming and establishment. *Trends in Cell Biology* 28(3):237–253. doi:10.1016/j.tcb.2017.10.008.

Yamashiro, C., K. Sasaki, S. Yokobayashi, Y. Kojima, M. Saitou. 2020. Generation of human oogonia from induced pluripotent stem cells in culture. *Nature Protocols* 15(4):1560–1583. doi:10.1038/s41596-020-0297-5.

Yamashiro, C., K. Sasaki, Y. Yabuta, Y. Kojima, T. Nakamura, I. Okamoto, S. Yokobayashi, Y. Murase, Y. Ishikura, K. Shirane, H. Sasaki, T. Yamamoto, and M. Saitou. 2018. Generation of human oogonia from induced pluripotent stem cells in vitro. *Science* 362(6412):356–360. doi:10.1126/science.aat1674.

Yatsenko, A. N., A. P. Georgiadis, A. Röpke, A. J. Berman, T. Jaffe, M. Olszewska, B. Westernströer, J. Sanfilippo, M. Kurpisz, A. Rajkovic, S. A. Yatsenko, S. Kliesch, S. Schlatt, and F. Tüttelmann. 2015. X-linked *TEX11* mutations, meiotic arrest, and azoospermia in infertile men. *New England Journal of Medicine* 372(22):2097–2107. doi:10.1056/NEJMoa1406192.

Yu, Y., T. C. Leete, D. A. Born, L. Young, L. A. Barrera, S.-J. Lee, H. A. Rees, G. Ciaramella and N. M. Gaudelli. 2020. Cytosine base editors with minimized unguided DNA and RNA off-target events and high on-target activity. *Nature Communications* 11:2052. doi:10.1038/s41467-020-15887-5.

Yuan, Y., L. Li, Q. Cheng, F. Diao, Q. Zeng, X. Yang, Y. Wu, H. Zhang, M. Huang, J. Chen, Q. Zhou, Y. Zhu, R. Hua, J. Tian, X. Wang, Z. Zhou, J. Hao, J. Yu, D. Hua, J. Liu, X. Guo, Q. Zhoug, and J. Sha. 2020. In vitro testicular organogenesis from human fetal gonads produces fertilization-competent spermatids. *Cell Research*. 30(3):244–255. doi:10.1038/s41422-020-0283-z.

Zanetti, B. F., D. P. A. F. Braga, A. S. Setti, R. C. S. Figueira, A. Iaconelli Jr., and E. Borges Jr. 2019. Preimblantation genetic testing for monogenic diseases: A Brazilian IVF centre experience. *Journal of the Brazilian Society of Assisted Reproduction* 23(2):99–105. doi:10.5935/1518-0557.20180076.

Zeng, Y., J. Li, G. Li, S. Huang, W. Yu, Y. Zhang, D. Chen, J. Chen, J. Liu, and X. Huang. 2018. Correction of the marfan syndrome pathogenic FBN1 mutation by base editing in human cells and heterozygous embryos. *Molecular Therapy: The Journal of the American Society of Gene Therapy* 26(11):2631–2637. doi:10.1016/j.ymthe.2018.08.007.

Zhang, M., C. Zhou, Y. Wei, C. Xu, H. Pan, W. Ying, Y. Sun, Y. Sun, Q. Xiao, N. Yao, W. Zhong, Y. Li, K. Wu, G. Yuan, S. Mitalipov, Z. Chen, and H. Yang. 2019. Human cleaving embryos enable robust homozygotic nucleotide substitutions by base editors. *Genome Biology* 20:101. doi:10.1186/s13059-019-1703-6.

Zhang, X. M., K. Wu, Y. Zheng, H. Zhao, J. Gao, Z. Hou, M. Zhang, J. Liao, J. Zhang, Y. Gao, Y. Li, L. Li, F. Tang, Z. J. Chen, and J. Li. 2020. In vitro expansion of human sperm through nuclear transfer. *Cell Research* 30:356–359. doi:10.1038/s41422-019-0265-1.

Zhao, D., J. Li, S. Li, X. Xin, M. Hu, M. A. Price, S. J. Rosser, C. Bi, and X. Zhang. 2020. Glycosylase base editors enable C-to-A and C-to-G base changes. *Nature Biotechnology.* doi:10.1038/s41587-020-0592-2.

Zhou, C., M. Zhang, Y. Wei, Y. Sun, Y. Sun, H. Pan, N. Yao, W. Zhong, Y. Li, W. Li, H. Yang, and Z. Chen. 2017. Highly efficient base editing in human tripronuclear zygotes. *Protein and Cell* 8:772–775. doi:10.1007/s13238-017-0459-6.

Zhou, F., R. Wang, P. Yuan, Y. Ren, Y. Mao, R. Li, Y. Lian, J. Li, L. Wen, L. Yan, J. Qiao, and F. Tang. 2019. Reconstituting the transcriptome and DNA methylome landscapes of human implantation. *Nature* 572:660–664. doi:10.1038/s41586-019-1500-0.

Zhou, Q., M. Wang, Y. Yuan, X. Wang, R. Fu, H. Wan, M. Xie, M. Liu, X. Guo, Y. Zheng, G. Feng, Q. Shi, X. Y. Zhao, J. Sha, and Q. Zhou. 2016. Complete meiosis from embryonic stem cell-derived germ cells in vitro. *Cell Stem Cell* 18(3):330–340. doi:10.1016/j.stem.2016.01.017.

Zhu, F., R. R. Nair, E. M. C. Fisher, and T. J. Cunningham. 2019. Humanising the mouse genome piece by piece. *Nature Communications* 10(1):1845. doi:10.1038/s41467-019-09716-7.

Zhu, P., H. Guo, Y. Ren, Y. Hou, J. Dong, R. Li, Y. Lian, X. Fan, B. Hu, Y. Gao, X. Wang, Y. Wei, P. Liu, J. Yan, X. Ren, P. Yuan, Y. Yuan, Z. Yan, L. Wen, L. Yan, J. Qiao, and F. Tang. 2018. Single-cell DNA methylome sequencing of human preimplantation embryos. *Nature Genetics* 50:12-19. doi:10.1038/s41588-017-0007-6.

Zlotogora, J. 1997. Dominance and homozygosity. *American Journal of Medical Genetics* 68:412–416.

Zuccaro, M. V., J. Xu, C. Mitchell, D. Marin, R. Zimmerman, B. Rana, E. Weinstein, R. T. King, M. Smith, S. H. Tsang, R. Goland, M. Jasin, R. Lobo, N. Treff, and D. Egli. 2020. Reading frame restoration at the EYS locus, and allele-specific chromosome removal after Cas9 cleavage in human embryos. *bioRxiv* 2020.06.17.149237; doi:10.1101/2020.06.17.149237.

A

The International Commission on the Clinical Use of Human Germline Genome Editing was tasked with developing a framework for scientists, clinicians, and regulatory authorities to consider when assessing potential clinical uses of human germline genome editing, should society conclude that heritable human genome editing (HHGE) applications are acceptable.

COMMISSION COMPOSITION

The U.S. National Academy of Medicine, the U.S. National Academy of Sciences, and the U.K.'s Royal Society appointed a Commission of 18 experts to undertake the Statement of Task. The Commission's membership spans 10 nations and 4 continents and includes experts in science, medicine, genetics, ethics, psychology, regulation, and law. Appendix B provides biographical information for each Commissioner.

In addition, an International Oversight Board (IOB) of leaders from national academies of sciences and international institutions was charged with ensuring that the Commission followed due processes, including approving the Statement of Task and membership of the Commission and ensuring that the Commission's report underwent rigorous external review prior to publication.

MEETINGS AND INFORMATION-GATHERING ACTIVITIES

The Commission deliberated from approximately June 2019 through March 2020 to conduct its assessment and prepare its final report. To address its task, the Commission analyzed information obtained from current literature and other publicly available resources and undertook information-gathering activities such as inviting stakeholders to share perspectives at public meetings, holding webinars, and soliciting public input online and in person.

Public Meetings and Webinars

Sessions at meetings and webinars held over the course of the study enabled Commissioners to obtain input from a range of stakeholders and members of the public.

The Commission's first meeting was held in August 2019 in Washington, DC. Public sessions provided an opportunity for the Commission to discuss its Statement of Task with the co-chairs of IOB and sponsoring organizations and to hear presentations on the state of understanding of genetics and genetic manipulation; on somatic genome editing translational pathways from scientists, developers, and regulatory bodies; and on the views of genetic disease patient communities.

In November 2019, the Commission held a second meeting and workshop in London, United Kingdom. The Commission heard from invited experts on topics such as the medical ethics of HHGE, how clinical use of HHGE would intersect with the use of assisted reproductive technologies, and technologies that might enable HHGE, including making and validating edits in embryos and germ cells, and what we can learn from animal models. In addition, the Commission hosted a session on governance developed in consultation with two members of the WHO Advisory Committee.

At a third meeting in January 2020, Commission members developed the conclusions and recommendations presented in this report.

In October 2019, the Commission also held a series of four public webinars on the state of research in relevant areas. These webinars covered (1) informed consent in the context of HHGE, (2) the impact of genome editing on embryo viability and the state of the science on editing spermatogonial stem cells, (3) homology directed repair and single cell genomics, and (4) validating on-target and off-target edits.

The list of speakers who provided input to the Commission in these meeting and webinar sessions is below.

Public Comments

The Commission's data-gathering meetings provided opportunities for the Commission to interact with a variety of stakeholders. Each public meeting included a public comment period, in which the Commission invited input from any interested party. The Commission also worked to make its activities as transparent and accessible as possible.

The study websites hosted by the U.S. National Academies and the U.K.'s Royal Society were updated regularly to reflect recent and planned Commission activities. Study outreach included a study-specific email address for comments and questions. A subscription to email updates was available to share further information and solicit additional comments and input to the Commission.

Live video streams with closed captioning were provided throughout the course of the study to allow the opportunity for input from those unable to attend public meetings in person. Information provided to the Commission from outside sources or through online comment is available by request through the National Academies' Public Access Records Office.

Call for Evidence

To inform its deliberations, the Commission invited responses to a public call for evidence during fall 2019. Several of the questions invited broad input on considerations associated with HHGE, while others asked for technical input in areas such as preclinical safety and efficacy and the use of genome editing in human embryos. Still other questions asked about considerations for informed consent, long-term monitoring, and oversight of HHGE.

There were 83 responses received. Respondents came from every continent and included academic leaders, lawyers, social scientists, philosophers, and representatives from disability advocacy groups, journals, national ethics councils, industry, and scientific societies.

Consulted Experts

The following individuals were invited speakers at data-gathering sessions of the Commission or provided other expert input.

Sonia Abdelhak
Institut Pasteur de Tunis, Tunisia

Britt Adamson
Princeton University, USA

Fabiana Arzuaga
Ministry of Science, Technology and Productive Innovation, Argentina

Richard Ashcroft
City University of London, U.K.

Christina Bergh
University of Gothenburg, Sweden

Peter Braude
King's College London, U.K.

Annelien Bredenoord
University Medical Centre Utrecht, Netherlands

Aravinda Chakravarti
New York University School of Medicine

Sarah Chan
University of Edinburgh, U.K.

Ellen Clayton
Vanderbilt University, USA

Chad Cowan
Harvard Stem Cell Institute, USA

James Lawford Davies
Hill Dickinson LLP, U.K.

Tarek El-Toukhy
Guy's and St Thomas' Hospital NHS Foundation Trust, U.K.

Frances Flinter
King's College London, U.K.

Denise Gavin
Food and Drug Administration, USA

Melissa Goldstein
George Washington University, USA

Margaret Hamburg
U.S. National Academy of Medicine and Co-chair of WHO Expert
    Advisory Committee, USA

Muntaser Ibrahim
University of Khartoum, Sudan

Pierre Jouannet
Paris Descartes University, France

Jin-Soo Kim
Seoul National University, South Korea

Robert Klitzman
Columbia University, USA

Bruce Levine
University of Pennsylvania, USA

Robin Lovell-Badge
Francis Crick Institute, U.K.

Sandy Macrae
Sangamo Therapeutics, USA

Julie Makani
Muhimbili University of Health and Allied Sciences, Tanzania

Nick Meade
Genetic Alliance U.K.

Shoukhrat Mitalipov
Oregon Health & Science University, USA

Vic Myer
Editas Medicine, USA

Kathy Niakan
Francis Crick Institute, U.K.

Sarah Norcross
Progress Educational Trust, U.K.

Helen O'Neill
University College London, U.K.

Kyle Orwig
Magee-Womens Research Institute, USA

Matthew Porteus
Stanford University, USA

Adam Pearson
Actor, Presenter, and Campaigner, U.K.

Catherine Racowsky
Brigham and Women's Hospital, USA

Jackie Leach Scully
University of New South Wales, Australia and Newcastle University, U.K.

Azim Surani
University of Cambridge, U.K.

Sarah Teichmann
Wellcome Sanger Institute, U.K.

Sharon Terry
Genetic Alliance, USA

Peter Thompson
Human Fertilisation and Embryology Authority, U.K.

Carrie Wolinetz
National Institutes of Health, USA

Xiaoliang Sunney Xie
Peking University, China

Hui Yang
Institute of Neuroscience, Chinese Academy of Sciences, China

Mohammed Zahir
Muhimbili University of Health and Allied Sciences, Tanzania

Commissioner Biographies

Kay E. Davies, DPhil., CBE, DBE, FMedSci, FRS, is professor of genetics in the Department of Physiology, Anatomy, and Genetics and associate head of development, impact, and equality in the Medical Sciences Division at the University of Oxford. She established the Medical Research Council (MRC) Functional Genomics Unit in 1999 and co-founded the Oxford Centre of Gene Function in 2000. She is co-director of the Muscular Dystrophy UK Oxford Neuromuscular Centre. Her research interests lie in the molecular analysis and development of treatments for human genetic diseases, particularly Duchenne muscular dystrophy, and the application of genomics for the analysis of neurological disorders and gene–environment interactions. She has published more than 400 papers and won numerous awards for her work. She co-founded Summit Therapeutics and Oxstem. Dr. Davies is a founding fellow of the Academy of Medical Sciences and was elected a fellow of the Royal Society in 2003. She was appointed governor of the Wellcome Trust in 2008 and was deputy chair from 2013 to 2017. In 2008 she was made Dame Commander of the British Empire for services to science.

Richard P. Lifton, M.D., Ph.D., NAS, NAM, is the 11th president of The Rockefeller University. His work uses human genetics and genomics to understand fundamental mechanisms underlying a wide range of human diseases. He is well known for his discovery that mutations with large effect on human blood pressure act by altering renal salt reabsorption, discoveries that have informed dietary guidelines and therapeutic strategies used worldwide to reduce blood pressure and prevent heart attacks and strokes, and for his development and use of exome sequencing for clinical diagnosis and disease gene discovery. Dr. Lifton graduated summa cum laude from Dartmouth College, obtained M.D. and Ph.D. degrees from Stanford University, and completed training in internal medicine at Brigham and Women's Hospital/Harvard Medical School. Prior to Rockefeller, he was chair of the Department of Genetics and Sterling Professor at Yale University, where he founded the Yale Center for Genome Analysis. He is a member of the National Academy of Sciences and the National Academy of Medicine and has served on the governing councils of both organizations. He currently serves on the scientific advisory boards of the Simons Foundation for Autism Research and the Chan-Zuckerberg Initiative Biohub and is a director of Roche and its subsidiary Genentech. He previously served on the advisory council to the director of the National Institutes of Health and co-chaired the National Institutes of Health Precision Medicine Working Group, which developed the plan for the "All of Us" Presidential Initiative. He has received numerous awards for his research, including the 2014 Breakthrough Prize in Life Sciences, the 2008 Wiley Prize, and the highest scientific awards of the American Heart Association, the American and International Societies of Nephrology, and the American and International Societies of Hypertension. He has received honorary doctorates from Northwestern University, Mount Sinai School of Medicine, and Yale University.

Hidenori Akutsu, M.D., Ph.D., is a director of the Department of Reproductive Medicine at the National Center for Child Health and Development in Tokyo, Japan. He is a member of the Expert Panel on Bioethics, Council for Science and Technology Innovation of Japan and a secretary of the Committee on Genome Editing Technology in Medical Sciences and Clinical Applications of the Science Council of Japan. His research explores mechanisms of preimplantation development and stem cell reprogramming, and he has derived human embryonic stem cells in Japan. Dr. Akutsu received his M.D. from Hirosaki University and completed his clinical training in obstetrics gynecology at Fukushima Medical University. He completed his Ph.D. at Fukushima Medical University School of Medicine.

Robert Califf, M.D., MACC, NAM, is the Donald F. Fortin, M.D., Professor of Cardiology at Duke University. He is also professor of medicine in the Division of Cardiology and remains a practicing cardiologist. Dr. Califf was the commissioner of food and drugs (2016–2017) and deputy commissioner for medical products and tobacco (2015–2016). Prior to joining the U. S. Food and Drug Administration (FDA), Dr. Califf was a professor of medicine and vice chancellor for clinical and translational research at Duke University. He also served as director of the Duke Translational Medicine

Institute and founding director of the Duke Clinical Research Institute. A nationally and internationally recognized expert in cardiovascular medicine, health outcomes research, health care quality, and clinical research, Dr. Califf has led many clinical trials and has more than 1,200 publications in the peer-reviewed literature. Dr. Califf has served on a number of advisory committees for the FDA and National Institutes of Health. He has led major initiatives aimed at improving methods and infrastructure for clinical research, including the Clinical Trials Transformation Initiative, a public–private partnership co-founded by the FDA and Duke.

Dana Carroll, Ph.D., NAS, is a distinguished professor in the Department of Biochemistry at the University of Utah School of Medicine. He was until recently interim director of the Public Impact Program at the Innovative Genomics Institute at the University of California, Berkeley. Dr. Carroll's research involves genome engineering using targetable nucleases. His laboratory pioneered the development of zinc-finger nucleases as gene targeting tools, and he continued working with the more recent Transcription activator–like effector nucleases (TALENs) and CRISPR-Cas nucleases, with much of the effort focused on optimizing the efficiency of these reagents for targeted mutagenesis and gene replacement. Dr. Carroll's current interests include the societal implications of genome editing. He received his Ph.D. from the University of California, Berkeley, and did postdoctoral research at the Beatson Institute for Cancer Research in Glasgow, Scotland, and at the Carnegie Institution Department of Embryology in Baltimore.

Susan Golombok, Ph.D., FBA, is professor of family research and director of the Centre for Family Research at the University of Cambridge, and was a visiting professor at Columbia University in New York in 2005–2006. She has pioneered research on the impact of new family forms on child development and is one of the world's leading experts on families created by assisted reproduction (i.e., in vitro fertilization, egg donation, sperm donation, and surrogacy). She has authored more than 300 academic papers and 7 books, and her award-winning research has contributed to policy and legislation on the family both nationally and internationally. She was a member of the U.K. government's surrogacy review committee in the late 1990s and a member of the Nuffield Council on Bioethics Working Party on Donor Conception in 2012–2013. Golombok is a specialist in longitudinal studies of children, an important element of the Commission's tasks of identifying ways to assess the balance between potential benefits and harms to a child produced by genome editing and of identifying and assessing mechanisms for long-term monitoring of children produced by genome editing. She received her Ph.D. from the University of London

Institute of Psychiatry in 1982 and was elected a fellow of the British Academy in 2019.

Andy Greenfield, Ph.D., has been a programme leader at the Medical Research Council's Harwell Institute since 1996, and his laboratory's research focuses on the molecular genetics of mammalian sexual development. From 2003 to 2007, Dr. Greenfield served as a member of the Wellcome Trust's *Molecules, Genes and Cells* Funding Committee, and from 2009 to 2018 he was a member of the U.K. Human Fertilisation and Embryology Authority. He chaired the authority's Licence Committee from 2014 to 2018 and was deputy chair of its Scientific and Clinical Advances Advisory Committee, for which he now serves an external advisor. In 2014 and 2016 Dr. Greenfield chaired two expert scientific panel reviews of mitochondrial donation techniques, important components of the regulatory process permitting mitochondrial replacement therapy in the United Kingdom. He has spoken on numerous occasions about the science and ethics of genomic technologies and their application in animals and humans. From 2014 to 2020, he was a member of the Nuffield Council on Bioethics and chaired its 2016 working group that reported on ethical issues surrounding the use of genome editing in a range of organisms and contexts. Dr. Greenfield graduated with a B.A. in natural sciences from St. John's College, University of Cambridge. He received his Ph.D. in molecular genetics from St. Mary's Hospital Medical School, Imperial College London, and was a postdoctoral fellow at the Institute for Molecular Bioscience, University of Queensland, Australia. He has an M.A. in philosophy from Birkbeck, University of London, and is a fellow of the Royal Society of Biology.

A. Rahman A. Jamal, M.D., Ph.D., MRCP, is the pro vice chancellor of the Universiti Kebangsaan Malaysia (UKM) Kuala Lumpur campus. He is also the founding director of the UKM Medical Molecular Biology Institute at the UKM, Kuala Lumpur, and a professor of pediatric oncology and hematology, and molecular biology. Dr. Jamal's research focus is on molecular biology of cancers, other non-communicable diseases, thalassemia, and rare diseases. He and his research team have discovered gene signatures associated with the pathogenesis of colorectal cancer, glioma, and leukemias. He has pioneered personalized and precision medicine at the UKM Medical Molecular Biology Institute and is now the chairman of the Task Force for Precision Medicine under the auspices of the Academy of Sciences Malaysia. Dr. Jamal is the principal investigator for the Malaysian Cohort project, is a member of the Asia Cohort Consortium and the International Health Cohort Consortium, and has been a member of The Wellcome Trust U.K. Grant Committee for Longitudinal Population Studies since 2018. He is the chairman of the National Committee for Ethics for Cell Research and

Therapy and a member of the National Committee for Clinical Research, both under the Ministry of Health Malaysia. Dr. Jamal is currently the project director for the UKM Specialist Children's Hospital, which will be the first dedicated hospital for pediatric patients in Malaysia. He graduated from UKM in medicine in 1985 and obtained his M.R.C.P. (pediatrics) from the Royal College of Physicians Ireland in 1991. He was awarded a Ph.D. in hematology and molecular biology in 1996 from the University of London and has a graduate diploma in healthcare leadership and management from the Singapore Management University.

Jeffrey Kahn, Ph.D., M.P.H., NAM, is the Andreas C. Dracopoulos Director of the Johns Hopkins Berman Institute of Bioethics, a position he assumed in 2016. He is the inaugural Robert Henry Levi and Ryda Hecht Levi Professor of Bioethics and Public Policy and professor in the Department of Health Policy and Management of the Johns Hopkins Bloomberg School of Public Health. Dr. Kahn works in a variety of areas of bioethics, exploring the intersection of ethics and health/science policy, including human and animal research ethics, public health, and ethical issues in emerging biomedical technologies. He has served on numerous state and federal advisory panels. He is currently chair of the National Academies of Sciences, Engineering, and Medicine's Board on Health Sciences Policy and has previously chaired its Committee on the Use of Chimpanzees in Biomedical and Behavioral Research (2011); Committee on Ethics Principles and Guidelines for Health Standards for Long Duration and Exploration Spaceflights (2014); and Committee on the Ethical, Social, and Policy Considerations of Mitochondrial Replacement Techniques (2016). He formerly served as a member of the National Institutes of Health's Recombinant DNA Advisory Committee. In addition to committee leadership and membership, Dr. Kahn is an elected member of the National Academy of Medicine and an elected fellow of The Hastings Center. He was the founding president of the Association of Bioethics Program Directors. Dr. Kahn's publications include 3 books and more than 125 scholarly and research articles. He speaks widely on a range of bioethics topics, in addition to frequent media outreach. From 1998 to 2002 he wrote the bi-weekly column "Ethics Matters" on CNN.com. Prior to joining the faculty at Johns Hopkins, Dr. Kahn was Maas Family Endowed Professor of Bioethics and director of the Center for Bioethics at the University of Minnesota. He received his M.P.H. from the Johns Hopkins Bloomberg School of Public Health and his Ph.D. from Georgetown University.

Bartha Maria Knoppers, J.D., Ph.D. (comparative medical law), is a full professor, Canada Research Chair in Law and Medicine, and director of the Centre of Genomics and Policy of the Faculty of Medicine at McGill

University. She was chair of the Ethics and Governance Committee of the International Cancer Genome Consortium from 2009 to 2017, has been chair of the Ethics Advisory Panel of the World Anti-Doping Agency since 2015, and has been co-chair of the Regulatory and Ethics Workstream of the Global Alliance for Genomics and Health since 2013. In 2015–2016, she was a member of the drafting group for the recommendation of the Organisation for Economic Cooperation and Development's Council on Health Data Governance and gave the Galton Lecture in November 2017. She holds four doctorates *honoris causa* and is a fellow of the American Association for the Advancement of Science, the Hastings Center (bio-ethics), the Canadian Academy Health Sciences, and the Royal Society of Canada. She is an officer of the Order of Canada and of Quebec, and was awarded the 2019 Henry G. Friesen International Prize in Health Research.

Eric S. Lander, Ph.D., NAS, NAM, is president and founding director of the Broad Institute of the Massachusetts Institute of Technology (MIT) and Harvard, and is professor of biology at MIT and professor of systems biology at Harvard Medical School. From 2009 to 2017 he also served as co-chair of the President's Council of Advisors on Science and Technology for President Barack Obama. A geneticist, molecular biologist, and mathematician, Dr. Lander has played a pioneering role in the reading, understanding, and biomedical application of the human genome. He was a principal leader of the Human Genome Project and has done pioneering work on mapping genes underlying human diseases and traits, human genetic variation, genome architecture, genome evolution, and genome-wide screens to discover the genes essential for biological processes using CRISPR-based genome editing. Dr. Lander has received numerous honors including the MacArthur Fellowship, the Breakthrough Prize in Life Sciences, the Albany Prize in Medicine and Biological Research, the Gairdner Foundation International Award (Canada), the Dan David Prize (Israel), the Mendel Medal of the Genetics Society (U.K.), the City of Medicine Award, the William Allan Award from the American Society of Human Genetics, the Abelson Prize from the American Association for the Advancement of Science (AAAS), the Award for Public Understanding of Science and Technology from the AAAS, the James R. Killian Jr. Faculty Achievement Award from MIT, and honorary doctorates from more than a dozen universities and colleges.

Jinsong Li, Ph.D., is a professor at the Shanghai Institute of Biochemistry and Cell Biology of the Chinese Academy of Sciences. Dr. Li's laboratory focuses on stem cells and embryonic development, and he has made fundamental contributions through his work in mice to the establishment of androgenetic haploid embryonic stem cells that can be used as sperm replacement to efficiently support full-term embryonic development upon injection into MII oocytes, leading to the generation of semi-cloned (SC) mice. Dr. Li has shown that this technology can be used as a unique tool for genetic analyses in mice, including medium-scale targeted screening of crucial genes or essential nucleotides of a specific gene involved in a developmental process; efficient generation of mouse models carrying defined point mutations related to human developmental defects; and one-step generation of mouse models that mimic multiple genetic defects in human diseases. Most recently, Dr. Li initiated a major genome tagging project to tag every protein in mice based on artificial spermatid–mediated SC technology, which may enable the precise description of protein expression and localization patterns, and protein–protein, protein–DNA, and protein–RNA interactions. Dr. Li received his Ph.D. from the Institute of Zoology of the Chinese Academy of Sciences in 2002 followed by postdoctoral training at The Rockefeller University.

Michèle Ramsay, Ph.D., is director and research chair of the Sydney Brenner Institute for Molecular Bioscience at the University of Witwatersrand, Johannesburg. The institute focuses on the development of new solutions to African health challenges by conducting biomedical molecular and genomic research. Dr. Ramsay's research interests include the genetic basis and molecular epidemiology of single-gene disorders in South African populations and the role of genetic and epigenetic variation in the molecular etiology of diseases and traits affected by lifestyle choices. She is a member of the Academy of Science of South Africa, immediate past president of the African Society of Human Genetics, and president of the International Federation of Human Genetics Societies. Dr. Ramsay received her Ph.D. in human molecular genetics from the University of Witwatersrand.

Julie Steffann, M.D., Ph.D., is director of the molecular genetics department at Necker-Enfants Malades Hospital in Paris and professor of genetics at the Paris University. She has run the Preimplantation Genetic Diagnosis Laboratory since 2003 and belongs to the mitochondrial diseases research team at the Imagine Institute in Paris. The Imagine Institute focuses on understanding the mechanisms of genetic diseases and inventing tomorrow's treatments for genetic diseases. Steffann conducts research on mitochondrial DNA disorders and their consequences on human early embryos. She investigates the potential impacts of mitochondrial DNA mutations on human embryo/foetal development and develops methods of prevention and treatment of mitochondrial DNA disorders. She received her M.D. in 2001 and her Ph.D. in Genetics in 2006 from the Paris-Descartes University.

B.K. Thelma, Ph.D., is a professor and J.C. Bose Fellow at the Department of Genetics at the University of Delhi. She has also served as a member of the Scientific Advisory Council to the Prime Minister of India from 2009 to 2014. From 2008, she has been a team leader of the Centre of Excellence on Genome Sciences and Predictive Medicine of the Department of Biotechnology, Government of India. Dr. Thelma has made original contributions in the field of human genetics and medical genomics. Her group has identified several novel disease causal genes for familial forms of Schizophrenia, Parkinson's disease, and intellectual disability. Her group has also been the pioneers in the identification of novel susceptibility conferring genes for rheumatoid arthritis and ulcerative colitis in the Indian population. Her current work focuses on Ayurgenomics, an innovative approach of combining the doctrines of Ayurveda, the Indian system of medicine for deep phenotyping of individuals, with contemporary genome analysis tools to address the phenotypic heterogeneity limiting our understanding of the genetics of common complex disorders; and functional genomics of rare genetic variants using cellular models of disease and CRISPR-based genome editing tools. In her persistent efforts to translate benefits of science to society, Dr.Thelma established early on the DNA-based diagnosis for fragile X syndrome and, more recently, newborn screening for inborn errors of metabolism to reduce the socio-economic burden of this large group of genetic disorders in the country. Dr. Thelma has been involved in a number of long-term follow-up studies and has contributed to several expert committees in areas of science and ethics. She received the Stree Shakti Science Samman award in 2012 and is a fellow of the Indian National Science Academy, Indian Academy of Sciences, and the National Academy of Sciences (India). Dr Thelma obtained her Master's degree in zoology from Bangalore University and received a Ph.D. in zoology from the University of Delhi.

Douglass Turnbull, M.D., Ph.D., FMedSci, FRS, is professor of neurology and director of the Wellcome Centre for Mitochondrial Research at Newcastle University. The Wellcome Centre for Mitochondrial Research focuses on understanding the clinical course of patients with mitochondrial disease and how this relates to the underlying disease mechanisms, identifying the molecular and genetic mechanisms causing mitochondrial disease, and developing techniques to prevent the transmission of mtDNA disease and improve treatment for patients with mitochondrial disease. Dr. Turnbull was also director of the MRC Centre for Ageing and Vitality, which is focused on understanding how aging mechanisms are influenced by lifestyle interventions and carries out studies aimed at promoting healthy aging. He was the lead for the National Health Service Highly Specialised Services for Rare Mitochondrial Services for Adults and Children. This service provides optimum care for patients with mitochondrial disease throughout the United

Kingdom with centres in Newcastle, London, and Oxford. This service was built on the back of clinical and basic research, and the Newcastle centre reviews more than 1,000 patients per year. The service has developed care pathways and patient guidelines that are used worldwide. Dr. Turnbull was elected a fellow of the Academy of Medical Sciences in 2004 and elected a fellow of the Royal Society in 2019. Dr. Turnbull received a knighthood in the Queen's Birthday Honours 2016 "for services to health care research and treatment, particularly mitochondrial disease." He received his bachelor of medicine, bachelor of surgery, M.D., and Ph.D. from Newcastle University.

Haoyi Wang, Ph.D., leads a research group in the State Key Laboratory of Stem Cell and Reproductive Biology at the Chinese Academy of Sciences' Institute of Zoology. The Wang laboratory focuses on developing novel technologies to achieve efficient and specific genome engineering, and applying them to study the function of genes and establish novel therapeutic methods. His laboratory has developed a zygote electroporation of nuclease method to generate genetically modified mouse models with high throughput and efficiency, the Casilio method to regulate gene transcription, and a method to generate CAR-T cells with multiplex gene editing. Dr. Wang previously worked on the development of a variety of genome engineering technologies, including a transposon-based "calling card" method for determining the genome-wide binding locations of transcription factors, TALEN-mediated genome editing in human pluripotent stem cells and mice, CRISPR-mediated multiplexed genome editing in mice, and CRISPR-mediated gene activation in human cells. Dr. Wang received his Ph.D. from Washington University in St. Louis.

Anna Wedell, M.D., Ph.D., is head of the Centre for Inherited Metabolic Diseases at Karolinska University Hospital and professor of medical genetics at the Karolinska Institute in Stockholm, Sweden. Dr. Wedell leads an integrated translational centre combining clinical and laboratory medicine, high-throughput genomics, and basic experimental science. The centre performs nationwide clinical diagnostics of inborn errors of metabolism, including the national neonatal screening program ("PKU test"). The centre also has a strong focus on mitochondrial medicine. Dr. Wedell is affiliated with the Science for Life Laboratory, a national infrastructure for high-throughput biology. She has implemented whole-genome sequencing into health care and has discovered a number of novel monogenic diseases. She received her M.D. in 1988 and her Ph.D. in medical genetics in 1994 from the Karolinska Institute. In 2006, she became board certified in clinical genetics after training at the Karolinska University Hospital. Dr. Wedell is a member of the Nobel Assembly at Karolinska Institutet and the Royal Swedish Academy of Sciences.

Glossary

Allele. A variant form of a gene at a particular locus on a chromosome. Different alleles can produce variations in inherited characteristics.

Androgenetic haploid embryonic stem cells (AG-haESCs). Cells derived from embryos generated either by injecting sperm into oocytes from which the maternal chromosomes have been removed, or by fertilizing eggs and removing the female pronucleus.

Aneuploidy. The presence of an abnormal number of chromosomes in a cell.

Assisted reproductive technology (ART). A fertility treatment or procedure that involves laboratory handling of gametes (eggs and sperm) or embryos. Examples of ART include in vitro fertilization and intracytoplasmic sperm injection.

Autosomal dominant. A pattern of inheritance in which an affected individual has one disease-causing copy of a gene and one copy of a gene with the non-disease-causing sequence, located on the autosomal chromosomes. The disease-causing copy of the gene determines the resultant phenotype. Humans have 22 pairs of autosomal chromosomes and 1 pair of sex chromosomes (see below).

Autosomal recessive. A pattern of inheritance in which an affected individual has a disease-causing sequence in both copies of a gene located on an autosomal chromosome. A single disease-causing copy of the gene is insufficient to cause the phenotype.

Blastocyst. A preimplantation embryo in placental mammals (occurring at about 5 days after fertilization in humans) having between 50 and 150 cells. The blastocyst consists of a sphere made up of an outer layer of cells (the trophectoderm), a fluid-filled cavity (the blastocoel or blastocyst cavity), and a cluster of cells in the interior (the inner cell mass). Cells from the inner cell mass, if grown in culture, can give rise to embryonic stem cell lines.

Cas9 (CRISPR-associated protein 9). A specialized enzyme known as a nuclease that has the ability to cut DNA sequences. Cas9 makes up part of the "toolkit" for the CRISPR-Cas9 method of genome editing.

Chromatin. The complex of DNA and proteins that forms chromosomes. Some of the proteins are structural, helping to organize and protect the DNA, while others are regulatory, acting to control whether genes are active or not and to promote DNA replication or repair.

Chromosome. A thread-like structure that contains a single length of DNA, usually carrying many hundreds of genes. This is packaged with proteins to form chromatin. The DNA within the complete set of chromosomes in each cell (23 pairs in humans) includes two copies of the genome, one from each parent. The chromosomes usually reside in the nucleus of a cell, except during cell division, when the nuclear membrane breaks down and the chromosomes become condensed and can be visualized as discrete entities.

Compound heterozygous. Having two different disease-causing alleles for the same disease.

CRISPR (clustered regularly interspaced short palindromic repeats). A naturally occurring mechanism found in bacteria that involves the retention of fragments of foreign DNA, providing the bacteria with some immunity to viruses. The system is sometimes referred to as CRISPR-Cas9 to denote the entire gene-editing platform in which RNA homologous with the targeted gene is combined with Cas9 (CRISPR-associated protein 9), which is a DNA-cutting enzyme (nuclease) to form the "toolkit" for the CRISPR-Cas9 method of genome editing.

Cultured cell. A cell maintained in a tissue culture allowing expansion of its numbers.

De novo. From the Latin, meaning "of new." As used in this report, describes mutations arising in the embryo that are not inherited from either parent.

Deoxyribonucleic acid (DNA). A two-stranded molecule, arranged as a double helix, that contains the genetic instructions used in the development, functioning, and reproduction of all known living organisms.

Diploid. Cells that contain a full set of DNA—half from each parent. In humans, diploid cells contain 46 chromosomes (in 23 pairs).

DNA cleavage. The process of introducing a double-strand break in DNA.

DNA sequencing. A laboratory technique used to determine the sequence of bases (A, C, G, and T) in a DNA molecule. The DNA base sequence carries the information that a cell needs to assemble protein and RNA molecules. DNA sequence information is important in investigating the functions of genes.

Dominant. A pattern of inheritance of a gene or trait in which, in a diploid cell, a single copy of a particular allele (a gene variant) confers a function independent of the nature of the second copy of the gene.

Double-strand break. A break in the DNA double helix in which both strands are cut, as distinct from a single-strand break or "nick."

Edit. A change to genomic DNA sequence (e.g., insertion, deletion, substitution) resulting from the application of genome editing components (e.g., nuclease, repair template).

Embryo. An animal in the early stages of growth and differentiation that are characterized by cleavage (cell division of the fertilized egg), differentiation of fundamental cell types and tissues, and the formation of primitive organs and organ systems. In humans, this stage extends from shortly after fertilization to the end of the eighth week after conception, after which stage it becomes known as a fetus.

Embryonic stem cell (ESC, also known as ES cell). A primitive (undifferentiated) cell from the embryo that has the potential to become a wide variety of specialized cell types (i.e., is pluripotent). It is a cultured cell derived from the inner cell mass of the blastocyst. An embryonic stem cell is not an embryo; by itself, it cannot produce the cell types, such as trophectoderm cells, necessary to give rise to a complete organism. Embryonic stem cells can be maintained as pluripotent cells in culture and induced to differentiate into many different cell types.

Enhancement. Improving a condition or trait beyond a typical or normal level.

Epigenetic effects. Changes to the chemical structure of DNA or the proteins that associate with DNA that can alter gene expression without changing the DNA sequence of a gene. For example, in the epigenetic phenomenon called genomic imprinting, molecules called methyl groups attach to DNA and alter gene expression according to parental origin.

Epigenome. A set of genome-wide chemical modifications to DNA and to proteins that bind to DNA in the chromosomes that affect whether and how genes are expressed.

Gamete. A reproductive cell (egg or sperm). Gametes are haploid (having only half the number of chromosomes found in somatic cells—23 in humans), and when two gametes unite at fertilization, the resulting one-cell embryo (the zygote) has the full number of chromosomes (46 in humans).

Gene. A functional unit of heredity that is a segment of DNA in a specific site on a chromosome. A gene typically directs the formation of a protein or RNA molecule.

Gene expression. The process by which RNA and proteins are made from the instructions encoded in genes. Gene expression is controlled by proteins and RNA molecules that bind to the genome or to the RNA copy and regulate their levels of production and the levels of their products. Alterations in gene expression change the functions of cells, tissues, organs, or whole organisms and sometimes result in observable characteristics associated with a particular gene.

Gene therapy. Introduction of exogenous genes into cells with the goal of ameliorating a disease condition. Can also be referred to as gene addition therapy.

Genetic variation. Differences in the sequence of DNA among people.

Genome. The complete set of DNA possessed by an organism. In humans, the genome is organized into 23 pairs of homologous chromosomes and comprises approximately 6 billion base pairs.

Genome editing. The process by which the genome sequence is changed through the intervention of a DNA break or other DNA modification.

Genome-wide association studies. A way for scientists to identify genes involved in human disease. A genome-wide association study involves searching the genome for small variations, called single-nucleotide polymorphisms (SNPs, pronounced "snips"), that occur more frequently in people who have a particular disease than in people who do not. Each study can look at hundreds or thousands of SNPs at the same time. Researchers use data from this type of study to pinpoint genes that may contribute to a person's risk of developing a given disease.

Genomics. The study of all the nucleotide sequences—including structural genes, regulatory sequences, and noncoding DNA segments—in the chromosomes of an organism or tissue sample.

Genotype. Genetic constitution of an individual organism or cell.

Germ cell. A sperm or egg cell.

Germline cell. A cell at any point in the lineage of cells that will give rise to a germ cell (see above). The germline is this lineage of cells. Eggs and sperm fuse during sexual reproduction to create an embryo, thus continuing the germline into the next generation.

Guide RNA (gRNA). In CRISPR systems, a small RNA that combines with a Cas protein to form the complex that cuts DNA. The gRNA contains a sequence of approximately 20 bases that specifies the target to be cut.

Haploid. Refers to a cell, usually a gamete or its immediate precursor, that has only 1 chromosome from each pair (a haploid cell in humans has a set of 23 chromosomes). In contrast, body cells (somatic cells) are diploid, having two sets of chromosomes (46 in humans).

Heritable genetic change. Modifications to genes that could be passed down through generations. While heritable human genome editing would involve using editing reagents with germline cells, not all such editing is intended to be inherited. There is a distinction between research that is conducted only in a laboratory and making genetic changes in a clinical setting to establish a pregnancy.

Heterozygous. Having two different variants (alleles) of a specific gene on the two homologous chromosomes of a cell or an organism.

Homologous. (Of genes) having a shared genetic sequence.

Homologous recombination. The recombining of two like DNA molecules, including a process by which gene targeting produces an alteration in a specific gene.

Homology-directed repair (HDR). A natural repair process used to repair broken DNA, which relies on a DNA "template" with homology to the broken stretch of DNA. This usually occurs during or after DNA synthesis, which provides this template.

Homozygous. Having the same variant (allele) of a specific gene on both homologous chromosomes of a cell or an organism.

Human Fertilisation and Embryology Authority (HFEA). The U.K.'s independent regulator overseeing the use of germ cells and embryos in fertility treatment and research. The Human Fertilisation and Embryology Act is the law under which the authority operates and which it upholds.

Implantation. The process by which an embryo becomes attached to the inside of the uterus (occurring at 7 to 14 days after fertilization in typical pregnancies).

In vitro. From the Latin, meaning "in glass." Pertains to procedures performed in a laboratory dish or test tube, or in an artificial environment.

In vitro fertilization (IVF). An assisted reproduction technique in which fertilization is accomplished outside of the body.

In vitro gametogenesis (IVG). The use of stem cells to generate male or female gametes.

In vivo. From the Latin, meaning "in the living." Pertains to procedures performed in a natural environment, usually in the body of the subject.

Indel. A sequence change caused by the insertion or deletion of DNA sequence.

Induced pluripotent stem cell (iPSC). A cell type induced by the introduction or activation of genes conferring pluripotency and stem cell–like properties. For example, cells already committed to a particular fate (e.g., skin) can be induced to become pluripotent. This is useful in regenerative medicine, where iPSCs can be introduced back into the donor of the original cells with much less risk of transplant rejection.

Institutional review board (IRB). An administrative body in an institution (e.g., a hospital or a university) established to protect the rights and welfare of human research participants who are recruited to participate in research activities conducted under the auspices of that institution. The IRB has the authority to approve, require modifications in, or disapprove research activities in its jurisdiction, as specified by both federal regulations and local institutional policy.

Intended edit. A planned change to the genomic DNA sequence at the target resulting from the application of genome editing components (e.g., nuclease, repair template).

Locus. (Of genes) The place where a gene is located on a chromosome.

Mitochondria. Small structures present in human cells that are the sites of important metabolic functions, including energy production.

Mitochondrial DNA (mtDNA). The genetic material contained within the mitochondria.

Mitochondrial replacement techniques (MRT). Treatment methods with the potential to reduce the transmission of abnormal mtDNA from a mother to her child, and thus avoid mitochondrial disease in the child and subsequent generations.

Monogenic disorder. A disorder that results from a mutation at a single genetic locus. A locus may be present on an autosome or on a sex chromosome, and it may be manifested in a dominant or a recessive mode. A monogenic disorder may also be referred to as a Mendelian disorder.

Mosaicism. Variation among cells, such that the cells are not all the same— for example, in an embryo when not all of the cells are edited.

Mutation. A change in a DNA sequence. Mutations can occur spontaneously during cell division or can be triggered by environmental stresses, such as sunlight, radiation, and chemicals.

Non-homologous end joining (NHEJ). A natural repair process used to join the two ends of a broken DNA strand back together. This process is prone to errors in which short DNA sequences are introduced into the strand of DNA.

Nuclease. An enzyme that can cut through DNA or RNA strands.

Off-target event (or off-target edit). When a genome editing nuclease alters a DNA sequence at a location other than the one to which it was targeted. This can occur because the off-target sequence is similar, but not identical to, the intended target sequence.

On-target event (or on-target edit). Editing of the DNA at a specified, targeted location in the genome.

Oocyte. A developing egg; usually a large and immobile cell.

Pathogenic variant. A genetic alteration that increases an individual's susceptibility or predisposition to a certain disease or disorder.

Penetrance. The proportion of people who have a particular genetic change (e.g., a mutation in a specific gene) and exhibit signs and symptoms of a genetic disorder. If some people who have the mutation do not develop features of the disorder, the condition is said to have reduced (or incomplete) penetrance.

Phenotype. Observable properties of an organism that are influenced by both its genotype and its environment.

Polygenic inheritance. A pattern of inheritance that occurs when one characteristic is controlled by two or more genes.

Preimplantation genetic testing (PGT). Involves checking the genes or chromosomes of early embryos for a specific genetic condition. During PGT, a single cell or a small number of cells is removed from the embryo at the eight-cell or blastocyst stage and DNA is isolated and genotyped by sensitive methods, such as the polymerase chain reaction.

Pronucleus. The haploid nucleus of an oocyte or sperm, either prior to fertilization or immediately after fertilization, before the sperm and egg nuclei have fused into a single diploid nucleus.

Recessive. A recessive allele of a gene is one whose effects are masked by the second allele present in a diploid cell or organism, which is referred to as dominant.

Recombinant DNA. A recombinant DNA molecule is made up of DNA sequences that have been artificially modified or joined together (recombined) so that the new genetic sequence differs from naturally occurring genetic material.

Repair template. A nucleic acid sequence used to direct cellular DNA repair pathways to incorporate specific DNA sequence changes at or near a target site.

Ribonucleic acid (RNA). A single-stranded molecule that transmits and regulates the DNA's instructions for the development, functioning, and reproduction of all known living organisms.

Sex chromosome. A type of chromosome that participates in sex determination. Humans and most other mammals have two sex chromosomes, X and Y. A female has two X chromosomes in each cell, while a male has an X chromosome and a Y chromosome in each cell.

Single nucleotide polymorphism (SNP). A variant DNA sequence in which the purine or pyrimidine base of a single nucleotide has been replaced by another such base.

Somatic cell. Any cell of a plant or animal other than a reproductive cell or its precursor. In Latin, "soma" means "body."

Spermatogonial stem cells (SSCs). The self-replicating precursors of sperm cells.

Target sequence. A nucleic acid sequence subject to intentional binding, modification, or cleavage. The alteration induced at the target site can be a "desired on-target event" or an "unwanted on-target event." The latter events are often due to non-homologous end joining (NHEJ)–mediated DNA repair processes.

Transcription. Making an RNA copy from a gene or other DNA sequence. Transcription is the first step in gene expression.

Transcription activator–like effector nuclease (TALEN). An artificial nuclease composed of an endodeoxyribonuclease fused to DNA-binding domains of transcription activator–like effectors (TALEs) that cleave DNA at a defined distance from TALE recognition sequences. For example, a TALEN may refer to a pair of TALE-FokI fusion proteins that must dimerize on opposite strands of DNA adjacent to a target site for cleavage.

Translation. The process of forming a protein molecule from information contained in a messenger RNA—a step in gene expression following transcription (the copying of RNA from DNA).

Translational pathway (clinical). The series of steps that a technology would need to go through to proceed from basic research to clinical use.

Tripronuclear embryos. Egg cells that are fertilized by two sperm cells instead of one, precluding them from developing into a fetus.

Trophectoderm. The outer layer of the developing blastocyst that will ultimately form the embryonic side of the placenta.

Unintended edit. A change to the genomic DNA sequence at a location distinct from the target sequence, which results from the application of genome editing components (e.g., nuclease, repair template).

Variant. Distinct forms of a gene present in a population that can differ somewhat in function, with some being advantageous to the organism and some being deleterious or neutral.

Vector. A vehicle that transfers a gene into a new site (analogous to insect vectors that transfer a virus or parasite into a new animal host). Vectors used in molecular cell biology and genetic engineering include plasmids and modified viruses engineered to carry and express genes of interest in target cells. The most clinically relevant viral vectors for gene transfer include retroviral, lentiviral, adenoviral, and adeno-associated viral vectors.

Whole-genome sequencing (WGS). A laboratory process that determines the complete DNA sequence of an organism's genome at a single time.

X-linked disease. A disease caused by a mutation in a gene on the X chromosome. The phenotype will be expressed in females who are homozygous for the gene mutation and in males. Females with just one copy of the mutated gene are carriers.

Zinc-finger nuclease (ZFN). A class of engineered enzymes including both a DNA-binding domain and a DNA-cleavage enzyme that can be used as a genome editing tool.

Zygote. The single, fertilized cell that results from the combination of parental gametes—the egg and sperm.

REFERENCES

HFEA (Human Fertilisation and Embryology Authority). 2014. *Third Scientific Review of the Safety and Efficacy of Methods to Avoid Mitochondrial Disease through Assisted Conception: 2014 Update.*

NASEM (National Academies of Sciences, Engineering, and Medicine). 2017a. *Human Genome Editing: Science, Ethics, and Governance.* Washington, DC: The National Academies Press.

NASEM. 2017b. *An Evidence Framework for Genetic Testing.* Washington, DC: The National Academies Press.

NCI (National Cancer Institute). NCI Dictionary of Genetics Terms. https://www.cancer.gov/publications/dictionaries/genetics-dictionary. Accessed July 24, 2020.

NHGRI (National Human Genome Research Institute). Talking Glossary of Genetic Terms. https://www.genome.gov/genetics-glossary. Accessed July 24, 2020.

NIST (National Institute of Standards and Technology). Genome Editing Consortium Lexicon. ISO/CD 5058-1 Biotechnology. Genome Editing. Part 1: Terminology (in development). https://www.iso.org/standard/80679.html.

Acronyms and Abbreviations

AG-haESCs	Androgenetic haploid embryonic stem cells
AIDS	Acquired immune deficiency syndrome
ARRIGE	Association for Responsible Research and Innovation in Genome Editing
ART	Assisted reproductive technology
CAR-T cells	Chimeric antigen receptor T cells
Cas	CRISPR associated protein
Cas9	CRISPR associated protein 9
CF	Cystic Fibrosis
CRISPR	Clustered regularly-interspaced short palindromic repeats
DNA	Deoxyribonucleic acid
ES cell	Embryonic stem cell
ESHRE	European Society of Human Reproduction and Embryology
FDA	Food and Drug Administration
FH	Familial Hypercholesterolemia
gRNA	Guide ribonucleic acid
HDR	Homology-directed repair
HFEA	Human Fertilisation and Embryology Authority

HHGE	Heritable human genome editing
HIV	Human immunodeficiency syndrome
hPGCLC	Human primordial germ-like cell
HSC	Hematopoietic stem cell
ICH	International Council for Harmonisation of Technical Requirements for Pharmaceuticals for Human Use
ICM	Inner cell mass
ICSI	Intracytoplasmic sperm injection
IFFS	International Federation of Fertility Societies
IOB	International Oversight Board
iPSC	Induced pluripotent stem cell
ISAP	International Scientific Advisory Panel
IVF	In vitro fertilization
IVG	In vitro gametogenesis
LDL	Low-density lipoprotein
MII	Metaphase II
MRT	Mitochondrial replacement techniques
MST	Maternal spindle transfer
mtDNA	Mitochondrial DNA
NGS	Next-generation sequencing
NHEJ	Non-homologous end joining
ntESC	Nuclear transfer embryonic stem cell
OHSS	Ovarian hyperstimulation syndrome
PAM	Protospacer-adjacent motif
PB	Polar body
PCR	Polymerase chain reaction
PGC	Primordial germ cell
PGCLC	Primordial germ cell-like cell
PGT	Preimplantation genetic testing
PNT	Pronuclear transfer
RNA	Ribonucleic acid
SART	Society for Assisted Reproductive Technology
SCD	Sickle cell disease
SNP	Single nucleotide polymorphism

SNV	Single nucleotide variant
SSC	Spermatogonial stem cell
TALEN	Transcription activator–like effector nuclease
WGS	Whole-genome sequencing
WHO	World Health Organization
ZFN	Zinc-finger nuclease

Acknowledgment of Reviewers

This report was reviewed in draft form by individuals chosen for their diverse perspectives and technical expertise. The purpose of this independent review is to provide candid and critical comments that will assist the institution in making its published report as sound as possible and to ensure that the report meets institutional standards for objectivity, evidence, and responsiveness to the study charge. The review comments and draft manuscript remain confidential to protect the integrity of the deliberative process. We wish to thank the following individuals for their review of this report:

Sonia Abdelhak, The Pasteur Institute of Tunis (Tunisia)
Ruth Chadwick, Cardiff University (United Kingdom)
Pat Clarke, European Disability Forum (European Union)
Bernd Gänsbacher, Technical University of Munich (Germany)
Mohammed Ghaly, Hamad Bin Khalifa University (Qatar)
Mary Herbert, Newcastle University (United Kingdom)
Rudolf Jaenisch, Massachusetts Institute of Technology (USA)
Kazuto Kato, Osaka University (Japan)
Tim Lewens, University of Cambridge (United Kingdom)
John Lim, Duke University-National University of Singapore (Singapore)
David Liu, Massachusetts Institute of Technology and Harvard University (USA)
Dennis Lo, The Chinese University of Hong Kong (Hong Kong)
Robin Lovell-Badge, The Francis Crick Institute (United Kingdom)
Luigi Naldini, Vita-Salute San Raffaele University (Italy)
Kathy Niakan, The Francis Crick Institute (United Kingdom)
Kelly Ormond, Stanford University (USA)
Sharon Terry, Genetic Alliance (USA)

Although the reviewers listed above provided many constructive comments and suggestions, they were not asked to endorse the conclusions or recommendations, nor did they see the final draft of the report before its release. The review of this report was overseen by **Suzanne Cory,** University of Melbourne (Australia), and **Janet Rossant,** The Gairdner Foundation (Canada). Acting on behalf of the study's International Oversight Board, they were responsible for making certain that an independent examination of this report was carried out in accordance with institutional procedures and that all review comments were carefully considered. Responsibility for the final content of this report rests entirely with the authoring Commission and the institutions.

NATIONAL ACADEMY OF MEDICINE

NATIONAL ACADEMY OF SCIENCES

THE
ROYAL
SOCIETY

The National Academy of Medicine (formerly the Institute of Medicine) was established in 1970 under the charter of the National Academy of Sciences to advise the nation on medical and health issues. Members are elected by their peers for distinguished contributions to medicine and health. Dr. Victor J. Dzau is president.

The National Academy of Sciences was established in 1863 by an Act of Congress, signed by President Lincoln, as a private, nongovernmental institution to advise the nation on issues related to science and technology. Members are elected by their peers for outstanding contributions to research. Dr. Marcia K. McNutt is president.

Learn more about the National Academy of Medicine and National Academy of Sciences at **www.nationalacademies.org**.

The **Royal Society** is a self-governing Fellowship of many of the world's most distinguished scientists. Its members are drawn from all areas of science, engineering, and medicine. It is the national academy of science in the United Kingdom. The Society's fundamental purpose, reflected in its founding Charters of the 1660s, is to recognize, promote, and support excellence in science and to encourage the development of use of science for the benefit of humanity.

Learn more about the Royal Society at **www.royalsociety.org**.

Consensus Study Reports document the evidence-based consensus on the study's statement of task by an authoring committee of experts. Reports typically include findings, conclusions, and recommendations based on information gathered by the committee and the committee's deliberations. Each report has been subjected to a rigorous and independent peer-review process and it represents the position of the authoring institutions on the statement of task.